CUTTING IT COSTS

Reduce unnecessary spending on information technology and align investments to higher value

PATRICK K. CHAPMAN

VOLUME 1: LARGE-SCALE CHANGE

VOLUME 2: OPERATIONAL OPPORTUNITIES

VOLUME 3: TACTICAL ACTIVITIES

Nordic Press
New York, NY

First Edition: January 2013

ISBN-10: 0985143029
ISBN-13: 978-0-9851430-2-2

This book is dedicated to KG Golshan, whose pragmatism, integrity, and decisiveness paved the way for progress in more private and public sector organizations than anyone will ever realize.

Contents

Acknowledgements

The author wishes to acknowledge Shilpa Arora and William Carleton for their support as well as their dedication to unmatched levels of professionalism and delivery quality. A special thanks to Jonathan Moore for his contributions to the subject matter addressed throughout this text. The author also wishes to acknowledge his children for their love, support, and incredible skill at making everyday a joy worth living.

Cutting IT Costs

Disclaimer

Before reading this collection of opportunities associated with reducing IT spending a few disclaimers are required:

1. The purpose of this text is to spark and circulate ideas, rather than provide detailed implementation guidance. If you're looking for detailed procedures for leveraging server virtualization for example, this isn't it. This text is focused on identifying concepts for cutting IT costs.

2. Every environment is different and what works for one organization both conceptually and how it's implemented, is frequently completely inapplicable to other organizations. Any actions provided are simply illustrative. Most sound implementations follow a standard analyze, design, implement, measure (or similar) approach; best judgment on applicability and approach is always required.

3. This book and its author makes no assurance or guarantee of savings of any level. All descriptions of concepts, benefits, savings, risks, and activities are based on historical observation of what other organizations have used to reduce or avoid unnecessary IT spending.

4. The information contained within this book is not, nor is it intended to be, legal advice. Consult an attorney for advice regarding any individual situation.

The author is happy to respond to any comments, feedback or questions and may be contacted via his LinkedIn profile at: www.linkedin.com/pub/patrick-chapman/9/a8a/b56/.

VOLUME 1 INTRODUCTION

LARGE-SCALE CHANGE

Last year organizations globally spent roughly 3.5 Trillion dollars (U.S) on the purchase, implementation, and operation of information technology (IT). By 2014 global spending on IT is expected to exceed 4 Trillion dollars. Over the course of the last decade organizations invested almost 20 Trillion dollars globally in information technology and information management, representing one of the single biggest areas of spending for organizations of almost any size and industry after human capital. Yet a quick search for books or articles about ways to reduce IT spending returns few results and the results one does receive are heavily focused on the acquisition of even more hardware and software.

One might assume this lack of coverage focused specifically on improving the efficiency of investments in IT is an indicator that organizations have already mastered the art of maximizing IT effectiveness while minimizing its cost. To be certain, the evolution and implementation of enterprise architecture concepts, governance models, and standardized service delivery frameworks represent significant efficiency levers that have driven down the cost of managing and delivering technical capability within and across many organizations. Surprisingly however, even these larger and well known cost levers are not embraced or fully implemented in many organizations including global conglomerates, government agencies, and even small to mid-size businesses. Other types of opportunities for reducing IT spending also frequently remain unrealized, with organizations relying instead on individual staff experience in IT and procurement to keep costs 'low'.

What makes reducing IT spending an often tricky endeavor is that it continues to represent one of the most important levers of value for most organizations. Indeed, virtually all organizations have advanced beyond simply leveraging technology to increase productivity and now consistently use technology to innovate products and services as well as find, grow, and service their customer or stakeholder base. Simply cutting spending in technology can have a crippling effect on the prospects of many organizations. Nevertheless, unnecessary spending on IT translates into, at a minimum, a missed opportunity to leverage the investment for greater returns in other areas. In many cases however, bloated and avoidable spending on IT represents an anchor that drags the entire organization down.

Given the criticality of technology to organizational health and opportunity, organizations must take care to operate with precision - to remove fat and avoid unnecessary spending without wounding important areas of technology investment. To cut wholesale, without understanding the levers of good versus bad IT spending, without understanding where efficiencies truly lie, is to operate on the organization's technical and operational footprint with a cleaver rather than the required scalpel. By understanding where opportunities to reduce IT spending exist, whether through the result of large-scale change or a series of smaller changes, individuals with even a modest span of control possess the capability to increase the efficiency of IT.

To provide greater visibility into the IT spending reduction mechanisms available to individuals and organizations, this text is divided into three separate volumes. The first volume, *Large-Scale Change*, focuses primarily on the major levers for reducing an organization's IT spending footprint by leveraging investments on an organization-wide basis (centralization). The second volume, *Operational Opportunities*, focuses to a large extent on opportunities associated with improving the way organizations continuously manage the maintenance and operation of technology. The third volume, *Tactical Activities*, focuses heavily on simple and immediate ways to reduce the acquisition cost of IT hardware, software, and associated support.

The ideas presented within this text are by no means an exhaustive list of opportunities to increase the efficiency and effectiveness of investments in IT. They simply represent an introduction to a series of activities, ranging from simple to complex, that are worth considering in a wide range of IT implementation and support organizations.

1

Common Drivers of Opportunity

Virtually every IT group possesses some level of opportunity to reduce IT spending without sacrificing technical capability, capacity, or quality. However, certain characteristics lend themselves to greater opportunities for increased efficiency. Organizations with significant opportunities for reducing undesirable IT spending frequently possess one or more of the characteristics summarized below.

Highly Decentralized IT Operating Environments

A fundamental characteristic of organizations possessing significant opportunity to reduce their level of investments in IT are those that use a highly disparate, segmented, or decentralized approach for providing IT services across the organization. An organization that uses multiple IT groups, each with different organizational structures, budgets, technologies, processes, and/or decision-makers across locations or sub-organizations is typically missing substantial opportunities to cut IT spending based on economies of scale, increased utilization of technical and human resources, and standardization.

Few Well-Defined and Documented Business and Technical Policies and Processes

Another major characteristic of inefficient IT organizations is a lack of well-defined policies and processes for IT acquisition, implementation, and management. If individuals or sub-organizations can approach IT in any manner they wish, including following their own security policies (or ignoring them), service levels, or development methods for example, then savings opportunities likely exist. Additionally, if these policies and processes are not enforced systematically as controls within supporting systems such as procurement or IT service management systems, the value of their existence on paper is often questionable based on the level to which they are actually adhered to.

A Lack of Centralized IT Spending Control

If the responsibility for acquisition of IT hardware, software, and services is dispersed across an organization or handled by individuals who do not specialize in IT procurement, it is highly likely significant opportunities for savings exist. A decentralized IT procurement approach often results in poor visibility into IT spending reducing an organization's opportunity to identify economies of scale or areas of duplication. The approach also frequently results in a lack of adherence to product and service standards (if they exist). In cases where general procurement resources are responsible for identifying sources and negotiating with IT vendors, significant opportunities often exist to reduce acquisition costs from the use of resources that specialize in IT acquisition processes and techniques (see Volume 3: Tactical Activities).

Reactive Approaches to Customer Needs
If IT resources always appear to be in fire-fighting mode, racing from one emerging requirement to the next, often relying on heroic efforts of one or a few resources to problem solve, there are likely opportunities for increased efficiency. Emergency requirements do come up, but the best IT organizations have well-defined paths of escalation for such requirements. Additionally, heavily reactive approaches do not typically embed the 'longer view', resulting in an increasing reliance on point solutions to address end-user requirements rather than leveraging more efficient enterprise-wide solutions and approaches.

A 'Unique Needs' Culture
If an organization is populated with various sub-organizations that contend they possess IT requirements that are different from other groups, efficiency opportunities are likely to exist in a variety of areas. An organization that has addressed a unique needs culture by employing separate IT groups or point solutions has frequently failed to embrace a number of core IT delivery concepts including 'classes of service' and platform standardization.

No Individual Possessing the Span of Control Necessary to Implement Efficient IT
In cases where the responsibility for all IT investments and delivery management do not ultimately roll up to a single individual, additional efficiencies are likely available to the organization. Of course this concept of unity of command has its limits, especially in loosely coupled operating structures, such as global conglomerates. However, in most organizations where no one individual possesses the ability to affect enterprise-wide management and change, there is

likely missed opportunity to increase resource leverage, reduce duplication, and pool acquisition volumes.

Low Levels of Technical Project Management Skills across the Organization

If technical projects are frequently behind schedule, over budget, or not delivering the intended benefits once implemented, money is being wasted. Perhaps second only to highly decentralized IT environments, poor technical project management skills account for some of the most significant waste events in IT organizations. Across the lifecycle, from planning issues such as failing to use existing standards or not collecting useful requirements, to implementation issues such as poor change tracking, a lack of solid technical project management skills can result in significant wastes of time and money for an organization.

Summary

In some cases, certain levels of decentralized control coupled with centralized planning and policy oversight, frequently referred to as a 'federated model', is an ideal blend of leveraged efficiency with end-user flexibility. However, the most efficient organizations have arrived at a federated model by design. In many cases, decentralized control and the resulting disparate approaches stem from a lack of planning and appropriate oversight. In these cases, organizations wishing to reduce IT spending by changing one or more of the characteristics described possess significant opportunities to not only reduce spending but potentially increase the effectiveness and quality of the IT capabilities delivered to the organization.

2

Good Versus Bad IT Spending

IT spending is frequently mischaracterized as an expense that should simply be minimized or avoided whenever possible, similar in view to that of office supplies or real estate square footage. In reality, IT spending is often one of the best areas of investment for an organization. Increased IT spending can, for example, significantly enhance employee productivity, provide a better experience for customers, help develop new products, or open new markets.

In many cases, organizations actually spend too little on IT relative to what they hope to accomplish, while simultaneously wasting too much. The dichotomy of needing to spend more while wasting too much is best explained by the concept of good versus bad IT spending. In reality, IT costs are driven by many factors. Most IT spending is desirable if it results in larger reductions in expenditures elsewhere or when it produces the desired level of returns or achievement of intended benefits (good spending). Some IT cost factors are relational such as headcount and for all intents and purposes cannot be directly controlled (neutral spending). However, some IT spending is simply undesirable when the expense is avoidable or could be reduced through greater efficiencies or other change activities (bad spending).

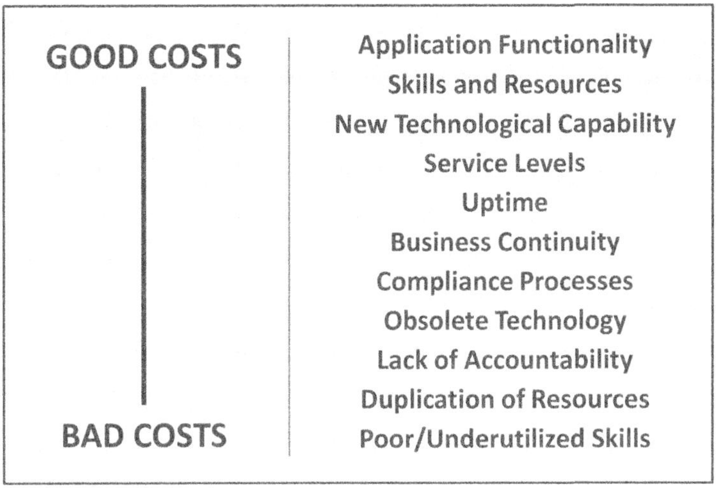

Figure 1: Because IT spending drives value, different sources of IT costs can represent either good spending or, if avoidable, bad spending.

Similar to the factors driving IT spending, activities associated with IT spending *reduction* may also be desirable or undesirable, depending on their effect on the nature and quality of IT service delivery.

Some cuts, for example, frequently result in little or no impact on technical capability or delivery quality. In other cases spending reductions may actually result in an equal or greater cost elsewhere. For example, cutting help desk hours can generate more costs associated with lost productivity than it saves.

Organizations with a poor understanding of good versus bad IT spending, little visibility into their spending, or poor spending controls frequently utilize undesirable savings levers due to a perceived lack of options, further weakening the organization's overall effectiveness.

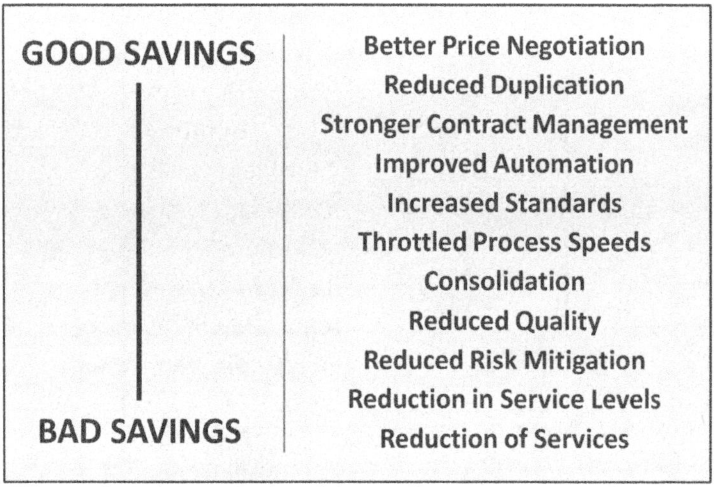

Figure 2: When reducing IT spending, some cuts of savings may be desirable if they reduce waste while others may be undesirable if they reduce capability or service levels.

The concept of good versus bad IT spending and good versus bad IT spending cuts is a foundational concept every individual undertaking IT efficiency initiatives needs to fully consider. Without awareness, the risk exists that reductions in IT spending will result in a reduction of capability and capacity when better opportunities to cut costs remain unrealized.

3

A Simple IT Spending Maturity Model

Traditional approaches for measuring organizational or project performance typically focus on the identification and collection of metrics or key performance indicators (KPIs). Improvements are then measured as changes in these metrics. While this approach is better than not identifying any performance measures, it typically falls significantly short in terms of measuring progress associated with IT spending reduction efforts.

The reasons for this are two-fold. First, IT cost cutting efforts are often about shifting IT investments to higher value, rather than cutting hard costs out of an IT budget. In these cases, the overall IT budget itself may not decrease, nor is a decrease perhaps even desirable; yet cutting unnecessary cost is the goal. Second, savings from improvements in IT efficiency often manifest themselves across many areas outside of IT, from customer service quality to individual employee productivity, itself often hard to measure for many types of functions or roles.

For these reasons, a basic IT spending maturity model as illustrated below represents a good, foundational mechanism to determine how efficient an IT organization is and what potential exists for efficiency. By first identifying, perhaps even quickly and somewhat subjectively, what the existing capability is from a maturity perspective, the organization can then define what it needs to accomplish to move itself to the next level.

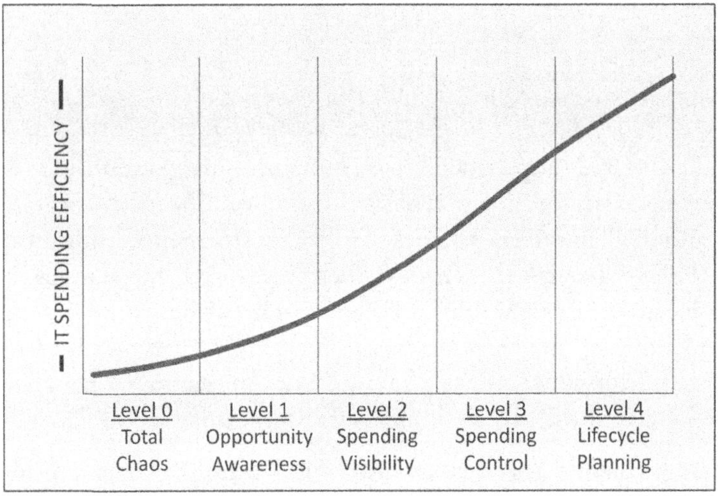

Figure 3: The efficiency of an organization's IT spending is measurable against a simple maturity model to better understand what level of additional efficiencies is feasible.

Level 0: Total Chaos

The organization possesses little or no awareness of its IT spending, does not contemplate its level of investment in IT, and leaves individual leaders to determine for their own sub-organizations how to best address its IT requirements.

Level 1: Opportunity Awareness
The organization recognizes that IT costs are driven by the outcome of planning and decisions made across the organization, but possesses little or no visibility into what the costs actually are and little or no control over the spending, except at the individual sub-organization level.

Level 2: Spending Visibility
The organization has implemented systems and processes to gain IT spending visibility by function, project, vendor, and product/service type to enable more effective decisions. The organization has begun to take advantage of smaller-scale opportunities to save such as obtaining volume discounts by pooling spending, but has yet to make large scale changes to the manner in which it manages its IT investments.

Level 3: Spending Control
The organization has implemented effective controls over IT spending through policies and processes enabling the organization to leverage a variety of economies of scale in the acquisition of technology and technical resources. The organization has also centralized a variety of foundational IT services such as user help desk support and data center resources and operations.

Level 4: Lifecycle Planning
In addition to IT acquisition controls and centralization of foundational services, the organization has centralized application development and integration activities and leverages standard methods for system lifecycle and infrastructure management. The organization is also making short, medium, and long term IT spending decisions based on a number of drivers such as business objectives, growth rates, project initiatives, and desired asset/resource utilization.

Summary

Simply put, organizations at lower maturity levels often exhibit higher than required spending to perform similar functions at the desired service levels. Organizations that achieve levels three or four are typically operating in an environment that minimizes the spending necessary to meet their requirements.

An additional benefit of using the IT spending maturity model as a lens for assessing the current state and potential for IT spend reduction is that it is a significantly easier story to tell and understand as efficiency efforts get underway. Occasionally, the desire for a more quantitative approach is driven by others to see the 'proof', or evidence that savings are real. The reality is the collection of quantitative savings metrics can be a rabbit hole from which otherwise valuable IT efficiency efforts are never able to climb out. With a maturity model approach, the business case for change is often easier to illustrate based on broader concepts.

The reality is in organizations or IT functions with a low level of overall maturity, anecdotal assessments of performance based on broad strokes are completely acceptable. Once maturity reaches level three or four, the collection of quantitative metrics may then make sense to drive further incremental value.

4

The IT Efficiency Phase Model

Another basic but beneficial framework used to evaluate an organization's potential for material reductions in IT spending is the IT Efficiency Phase Model (IT-EPM). The IT-EPM illustrates the required levels of organizational change and disruption relative to the expected achievement of efficiency in IT service delivery.

The level of change required to achieve a desired efficiency represents a critical concept for organizations wishing to cut costs. In many organizations, the scale of change acts as the single biggest obstacle for realizing savings. Whether from overcoming simple inertia or addressing real fears of job loss, change management needs often impede progress significantly greater than simply not knowing which efficiency levers to pull.

For individuals wishing to reduce IT spending, understanding the level of change the organization is likely capable of digesting as well as their own realistic span of control is helpful in knowing whether to pursue larger scale opportunities or focus on smaller tactical activities.

Put simply, smaller scale efforts will result in smaller scale savings. Larger changes, such as organizational consolidation, typically deliver significantly greater levels of efficiency. Ultimately, given the law of diminishing returns, some efforts may simply cost more to implement than the savings they deliver.

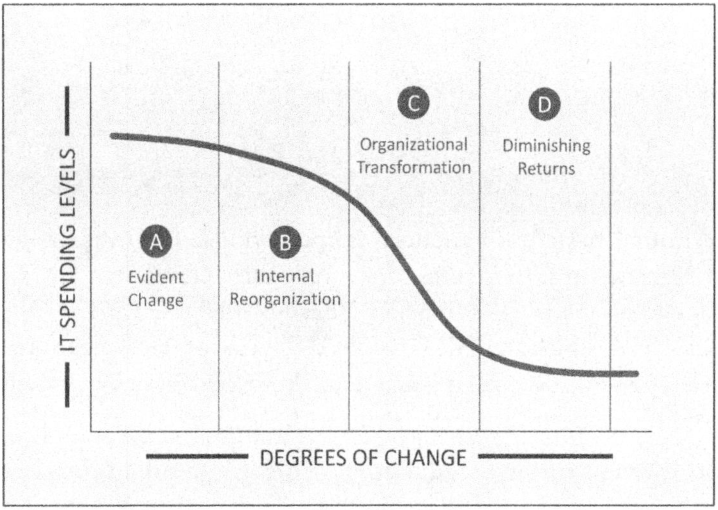

Figure 4: The degree to which IT efficiencies can be achieved in an organization is frequently a function of the level of change the organization is willing to engage in.

A. Evident Change

At one end of the spectrum, an organization that has not previously contemplated ways to reduce unnecessary IT spending will often find opportunities to implement obvious, low impact, short term changes that result in immediate savings. The result of these activities is frequently some level of spend reduction with minimal or no internal or external organizational impact.

B. Internal Reorganization
To achieve broader and more material savings, an IT organization, or more frequently multiple internal IT organizations, may undergo substantial realignment and reorganization of a variety of IT functions resulting in a more efficient centralized and standardized operating environment. The result of these types of changes is frequently significant internal impact relative to the IT organization(s) coupled with material IT spending reductions in the mid to long-term time horizon.

C. Business Transformation
Given the high level of dependency between IT spending and broader organizational operations, substantial business transformation across an entire organization typically results in the largest longer term reductions in IT spending, assuming the transformation includes centralization of IT functions as well. Changing the way the entire organization is structured in order to leverage a common, optimized set of systems and processes, frequently results in substantial organizational flux during the transition to a different operating model. These types of transformation events are frequently driven by crises or other large-scale restructuring efforts such as post-merger integration.

D. Diminishing Returns
In some cases, the level of control or application of resources can exceed the benefits associated with eliminating every potential redundancy or addressing every exception. The result of efforts to achieve 100% compliance or standardization can result in additional overhead driving net increases in IT spending.

Summary
As the model illustrates, there are likely no 'silver bullets' that enable an IT organization to magically reduce its

spending significantly without wholesale changes in its organizational and delivery approach. This is not to say that fully centralized and highly automated organizations do not possess any opportunities for savings. However, in the case of organizations that have already achieved a very high level of IT spending maturity, material spending reduction will likely come as the result of trade-offs in IT capabilities or capacity.

The IT-EPM also indirectly illustrates another common reality - organizations often need to begin with smaller scale spending reduction activities before the organization and its people are ready and willing to embrace larger change efforts that lead to material savings. Whether it's the need to demonstrate success before tackling bigger spend reduction initiatives or the need to illustrate that material savings won't come without it, the initiation of one or several smaller scale efforts is often exactly the right approach for pushing forward bigger IT spending reduction initiatives.

5

Evaluating Savings Opportunities

When evaluating potential opportunities to cut IT costs, there are a number of factors to consider. Of obvious concern is whether the level of time and money required to realize savings is worth the investment. This straightforward evaluation is based simply on whether the savings will exceed the required investment as well as any other opportunities that compete for the same time and money.

In many cases however, the specific levels of investment required for any given (or combined) IT efficiency effort are difficult to accurately calculate until the effort is initiated and some progress is made. Similarly, the savings themselves may be difficult to even estimate until some level of assessment or review is performed. Given this reality, it is often helpful to look initially instead at other attributes that indicate general levels of expected investment and return. In the case of IT savings opportunities, typical dimensions impacting investment and savings levels include implementation complexity, general savings potential, the savings horizon, and whether the savings stem from a single event or delivered continuously, as summarized in the figure below.

Implementation Complexity	
High	Significant time/$ investment and/or organizational change
Medium	Some time/$ investment and/or organizational change
Low	Minimal time/$ investment and/or organizational change
Savings Potential	
Very High	Greater than $2.5M
High	Between $1.5M - $2.5M
Medium	Between $500K - $1.5M
Low	Less than $500K
Savings Horizon	
Long-term	Savings realized after 18 to 24 months or longer
Mid-term	Savings realized within 6 to 18 months
Short-term	Savings realized within 0 to 6 months
Savings Frequency	
Continuous	Savings realized on a continuous basis
One-time	Savings realized one time

Figure 5: Common attributes for assisting in the evaluation and selection of IT savings opportunities.

Implementation Complexity

The complexity of implementing the changes necessary to realize savings represents a significant dimension impacting the decision whether to pursue a particular opportunity to cut IT costs. A simple framework for examining implementation complexity is established by looking at this dimension as three potential levels:

- **High** - The effort will require significant levels of internal resource time, substantial monetary investment, and/or substantial organizational changes to its structure or culture.

- **Medium** - The effort will require some internal resource time, some monetary investment, and/or some level of organizational changes to realize savings.
- **Low** - The effort will require little or no resource time, monetary investment, and low or no levels of impact on the organization's structure or culture.

Implied in these general levels of complexity are the associated levels of risk that accompany increasing complex initiatives. Potential issues such as sustained sponsorship commitment or continued availability of resources and funding will greatly impact the success of increasingly sophisticated and lengthy efforts and must be weighed accordingly when evaluating opportunities.

Savings Potential

Identifying highly accurate quantitative savings estimates is often almost impossible for IT cost cutting initiatives unless the scope of evaluation is extremely narrow. For example, estimating savings from using one versus two servers for a particular project is fairly straightforward to quantify. However, the savings from consolidating a large data center into fewer numbers of servers is significantly more complex based on a wide range of potential technical limitations that won't be uncovered until some level of review and effort is underway. In cases such as this, using a scale of ranges is helpful:

- **Very High** - Potential savings greater than $2.5M.
- **High** - Potential savings between $1.5M to $2.5M.

- **Medium** - Potential Savings between $500K to $1.5M.
- **Low** - Potential savings of less than $500K.

The number ranges are generalized and will likely change up or down depending on the size and scale of a given organization's annual IT budget. However, for any organization, even opportunities with low savings potential may be ideal, or indeed high priority, if they are simple to realize and recurring in nature.

Savings Horizon

Of often equal importance to savings potential, is the time frame under which the savings would likely be realized. A simple set of high-level categories for an opportunity's savings horizon includes:

- **Longer Term** - Savings are likely realized after 18 to 24 months from initiation of the effort.
- **Mid-Term** - Savings are likely realized in 6 to 18 months from initiation of the effort.
- **Shorter Term** - Savings are likely realized in 0 to 6 months from initiation of the effort.

Depending on the goals and objectives of any given organization, savings horizon may be a critical selection factor. For example, many organizations begin to focus heavily on IT efficiency not because IT isn't delivering its desired value, but because other factors are emerging, such as constrained budgets, falling revenues, or cash flow issues.

Savings Frequency

Another important consideration is whether the savings from a particular opportunity will be continuous or result only in a one-time savings. For example, consolidating several call centers into a single call center might result in a continuous savings of several million per year over the previous approach. However, an invoice audit which finds several thousand in recoverable overpayments will only save the thousands that one time.

- **Continuous** - Savings are realized on a continuous basis
- **One-Time** - Savings are realized as a single event

The frequency of savings represents a vital attribute as even small savings, when continuously realized, may represent an equal or better opportunity than a larger opportunity that may only result in a one-time savings.

Summary

By using a general framework for evaluating one or a series of potential opportunities in the absence of more quantitative data, decision making associated with which opportunity to pursue and in what priority is greatly facilitated. As illustrated in Figure 6 on the following page, the approach can help present to organizational leadership which opportunities to pursue based on a variety of internal needs. If for example the desire is immediate savings, the organization will likely pursue lower level complexity, shorter horizon initiatives even if they deliver lower overall savings.

Not described in this approach is the concept of soft savings, such as the indirect benefits of being able to apply resources and technologies made available through improved efficiency to enhance performance in other areas. In some cases it may be beneficial to highlight any identified soft benefits in addition to desired hard savings to further boost the business case of one or more potential initiatives.

Opportunity	Implementation Complexity	Savings Potential	Savings Horizon	Savings Frequency
Improve Requisition Function	Low	High	Long-term	Continuous
Implement Formal Governance Function	High	Very High	Long-term	Continuous
Improve Service De-Provisioning	Medium	Medium	Short-term	Continuous
Improve Billing Reconciliation	Medium	Low	Short-term	Continuous
Consolidate 2 Help Desks	Medium	High	Mid-term	Continuous
Delay Desktop Refresh	Low	Medium	Mid-term	One-Time
Improve Contract Oversight	High	Low	Long-term	Continuous

Figure 6: Example evaluation of several efficiency opportunities using common attributes.

If this simple evaluation model has generated in the reader's mind a number of other valuable assessment dimensions that would be helpful for evaluating opportunities, perfect! For any given organization, the importance of these and other opportunity attributes vary. The goal is to leverage a straightforward approach for assessing and prioritizing opportunities within a given organization in the absence of hard data or a desire to avoid significant time and investment in detailed analysis and estimation.

6

Centralizing IT Management and Operations

One of the most significant drivers of IT efficiency is the level of centralization of IT functions achieved within an organization. At its core, functional centralization is about minimizing duplicative investments and activities and maximizing the utilization of investments in technology and the resources supporting it.

The concept of centralization is based on continuous functional control and oversight. Centralization is not the same as consolidation, which is an activity, not a management approach. Consolidation activities however often result from transitions to a centralized approach, as duplicative investments are shed or stopped. Consolidation is the primary activity driving IT cost savings from transitioning to a centralized approach while ongoing centralized control is the activity that avoids future unnecessary spending.

The concept of centralized IT is also different from geographic consolidation. Often times a transition to centralized IT approaches is confused as a geographical consolidation, with remote resources who once governed local IT fearful that the need for local support will be terminated. However, technology and its accompanying support frequently must exist wherever resources requiring it exist.

To realize the benefits of centralized IT management, organizations typically migrate from some degree of decentralization. This process frequently manifests itself as a transition from independent or semi-independent IT groups, segmented across one or more sub-organizations, into a centralized functional orientation as illustrated below.

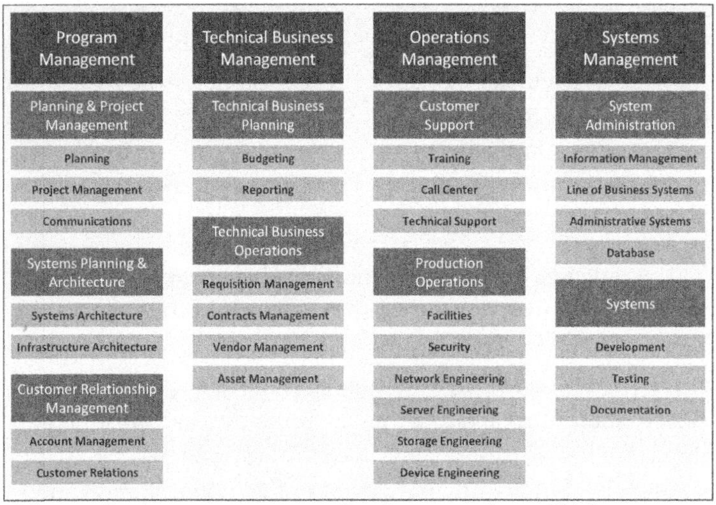

Figure 7: A common centralized approach to IT focuses on grouping activities using a functional framework.

Additionally, given a functional orientation and the removal of sub-organization specific groups, new or enhanced planning and technical business functions are frequently required to address gaps created by the realignment.

A typical functional IT organizational structure often consists of four core functions: program management; technical business management; operations management; and applications/systems management. An IT organization based on this functional approach is typically structured and populated with personnel responsible for specific functions. However, the functional model presented here is not intended to represent a single ideal organizational structure and is instead illustrative of the functions the proposed organization would fulfill.

Provided below is a summary of the activities each centralized function would conduct in order to meet the needs of the broader organization. However, in many organizations, it's not ideal to fully centralize some IT functions. In these cases, the goal is to determine to what degree, if any, a particular function should be centralized - from simple policy or process alignment to full integration of all staff and management into a single operating unit. The goal, as always, is striking an appropriate balance between efficiency and meeting the needs of the organization.

Program Management

A centralized program management function represents the primary control mechanisms responsible for determining the direction of IT. To accomplish this, the program management function typically consists of at least three sub-functions: planning and project management; systems planning and architecture; and customer relationship management.

Planning and Project Management

The planning and project management sub-function is responsible for establishing high level, longer term IT capability and investment targets for the organization as well as establishing and ensuring adherence to program and project management policies and processes. The sub-function's major focuses include:

- **Planning** - Developing strategic IT plans/roadmaps and enterprise-wide policies; coordinating enterprise architecture development and maintenance; reviewing investment requests; preparing chartering packages; managing IT steering/governance processes; maintaining systems portfolio information; establishing the overall technological direction of the organization.
- **Project management** - Providing IT project management support to help ensure adherence to budget, schedule, and quality, including the use of Earned Value Management (EVM) where appropriate.
- **Communications** - Communicating management's aims and directions to its stakeholders; developing communication plans; maintaining customer contact channels; developing communication policies.

Systems Planning and Architecture

The systems planning and architecture sub-function is essential for translating the organization's IT plans and direction into a standardized framework for the deployment of systems and supporting technologies. This sub-function's major activities encompass:

- **Systems architecture** - Defining the parameters associated with how systems will be developed or integrated; selecting core system/application component standards and platforms; selecting standard integration mechanisms; ensuring system/application solutions adhere to standards and specified architectures.

- **Infrastructure architecture** - Defining an overall enterprise architecture that is common, or shared, across the enterprise, as opposed to being specific to particular organizational units; selecting core processing, storage, networking, and other technical platform standards; ensuring technical solutions adhere to standards and specified architectures.

Customer Relationship Management

The customer relationship management sub-function focuses on helping line of business organizations translate business goals and requirements into technical requirements as well as making subsequent investments in specific technologies. It is not uncommon for many organizations to minimize or fail to invest adequately in this area, resulting in poor realization of business benefits from deployed technologies, despite seemingly adequate governance and steering functions. This is the nature of technology deployment - a system designed and implemented to automate a process for example often proves to do anything but once implemented, due to a lack of continuous business and technical collaboration throughout the process. Thus the primary focuses of customer relationship management are designed to mitigate this risk as well as enhance technology's value to the organization. Specific activities include:

- **Account management** - Assisting sub-organizations and/or lines of business in meeting essential IT objectives before, during, and after solutions implementation. This includes helping to identify the need for automated systems; helping define those systems and serving as the primary contact between the IT organization and its customers during the lifecycle of a solution including requirements definition, solution market analysis, and product selection; ensuring solutions comply with appropriate architectures; providing support in developing charter documents.

- **Customer relations** - Enhancing or maintaining the value and benefits of the product or service after implementation; escalating general IT issues; handling policy or service level exceptions.

Technical Business Management

The technical business management function represents a critical area of centralized activity providing an organization with the required levels of spending visibility while managing a number of important acquisition and asset management processes. The technical business management function typically consists of two major sub-functions: technical business planning and technical business operations.

Technical Business Planning

Technical business planning provides mechanisms for coordinating the development and execution of IT budgets through two primary areas of activity:

- **Budgeting** - Establishing the IT budget including preparation and review of budget submissions to finance organization(s); managing the process of activity and contract funding.

- **Reporting** - Assuring that reporting requirements are met, preparing reports, providing and maintaining mechanisms for reporting.

Technical Business Operations

The technical business operations sub-function includes a range of support areas focused on managing the acquisition and subsequent asset lifecycle associated with the organization's IT investments. Specific areas of activities include:

- **Requisition management** - Monitoring the policy and process by which goods or services are requested; preparing acquisition documents; selecting procurement channels; purchasing goods or services or coordinating purchases with acquisition group(s).

- **Contracts management** - Maintaining the appropriate mix of contract vehicles to address the organization's needs; performing administrative contract tasks and activities.

- **Vendor management** - Establishing processes for management and monitoring of vendors to make sure that they are meeting agreed to performance metrics and taking appropriate actions when they do not; working with key suppliers and vendors to ensure quality, alignment, and value.

- **Asset management** - Managing demand and guiding acquisition; deploying and disposing of assets to make the most of their service delivery potential; managing asset risks and costs over their entire lifecycle.

Operations Management

The operations management function includes all direct user support activities as well as a broad range of activities associated with maintaining the infrastructure required to delivery IT capabilities to the organization. In organizations possessing highly decentralized IT functions, these areas often represent the best and first areas of centralization and consolidation, delivering substantial efficiencies as a result.

Customer Support

Customer support, frequently referred to as 'seat management' activities, typically encompasses a number of user-oriented support areas including:

- **Training** - Developing and delivering education and training programs to assure users gain all available value from IT investments.
- **Call Center** - Fielding incoming help and service requests; handling 1st support tier issues; managing routing and supporting assignment of tickets from initiation to closure.
- **Technical/desk-side support** - Providing hands-on support to handle 1st and 2nd support tier issues; ensuring proper escalation to additional support tiers; working with vendors to resolve issues; ensuring appropriate use of support contract vehicles.

Production Operations

Production operations includes a number of activities focused on managing the availability and performance of computing and network resources upon which the delivery of systems and other network resources rely. Centralized planning, implementation, and delivery of these functional areas often drive the largest efficiencies within an organization while resulting in minimal, if any, impact on an organization's users. Production operations functions typically include:

- **Facilities** - Providing a suitable environment and physical controls that protect IT investments and supporting personnel or users.

- **Security** - Safeguard IT investments and information against unauthorized use, disclosure, modification, damage or loss; complete security compliance activities.
- **Network engineering** - Defining networking service and security requirements; acquiring, implementing, and configuring hardware and software to meet network requirements; supporting implemented networking solutions.
- **Server engineering** - Determining current and forecasted processing needs; acquiring and implementing hardware and software to meet requirements; supporting implemented computing solutions.
- **Storage engineering** - Determining current and forecasted storage requirements; acquiring, implementing and configuring hardware and software to meet requirements; supporting implemented storage solutions.
- **Desktop engineering** - Determining desktop computing needs; defining deployment/product tiers; acquiring and implementing hardware and software to meet engineering and support requirements.

Systems Management

The systems management function encompasses maintenance of an organization's existing systems and information repositories as well as the development and implementation of new or enhanced system capabilities. Centralization of these areas often represent a subsequent initiative for increasing efficiency, after centralization of operations management functions. The two primary sub-functions of systems management include system administration and system development.

System Administration

System administration activities focus on helping an organization maintain the performance, integrity, and access control associated with the organizations various systems and often encompasses:

- **Information management** - Managing applications and systems performing information management functions such as documented or content management.
- **Line of Business/Mission Systems** - Managing applications and systems performing line of business or mission-oriented functions.
- **Administrative Systems** - Managing applications and systems performing administrative functions.
- **Databases** - Managing shared database platforms.

System Development

The system development sub-function, while straightforward in its general focus, is an area where organizations frequently experience their largest IT waste 'events'. It is not uncommon for organizations to abandon or change the direction of development or integration efforts to such an extent that any existing investment in time and money is unrecoverable or wasted. To mitigate these risks, organizations are increasing their use of agile or highly iterative development approaches, as well as implementing stronger customer relationship management functions in order to identify required adjustments in technology earlier in the process. The primary activities of a centralized systems development function include:

- **Development** - Determining requirements for, designing, and building or integrating systems/applications.
- **Testing** - Performing usability and other testing of new and altered systems/applications.
- **Documentation** - Preparing the documentation required by the system's development lifecycle policies.

Critical Success Factors

Personnel within a functionally centralized structure typically focus on specific areas of activity for the entire organization as opposed to specific sub-organizations. However, individuals within a specific function may exclusively focus on providing that function for a single sub-organization if it is determined that the best underlying structure for a function is to have individuals within that function be 'customer-specific'. This does not affect the overall centralized approach and instead simply represents a way a particular function's workload is distributed.

It is also important to re-emphasize that the described functional approach does not represent an organizational chart. Significant planning, analysis, and design efforts are typically necessary to develop a functionally-focused organizational chart that embodies the required centralized capabilities while maintaining or increasing end-user touch points.

Additionally, outside support vendors or contractors often play an important part in both the transformation of an IT organization as well as its ongoing operations. A critical component of success is restructuring contracts for services so they too are functionally oriented rather than location or sub-organization oriented. Any new statements of work governing contractor resources should also embed and adhere to the policies and processes established for associated functions.

Line of business support should also receive increased treatment from a functional perspective relative to other areas of support, such as administrative systems. Existing line of business systems and support activities may need additional flexibility or support, where standardization at the appropriate levels is best balanced against careful consideration of business needs.

Other Considerations

A frequent concern in transitioning to a centralized organizational approach to IT management is how to get to the end state with the minimal amount of operational disruption. To minimize disruption to the organization a number of potential activities are often considered, including:

1. The establishment of a new IT organization at the top level of the organizational structure, if one does not already exist. Organizational placement outside line of business organizations is a critical step in establishing a centralized approach. While it is possible to simply take the largest line of business IT organization that exists within an organization and increase its responsibilities without moving its organizational placement, such approaches have proven to create a number of transition and delivery issues that in some cases cause other organizations to resist centralization efforts.

2. The new IT organization will need to appoint a single Chief Information Officer (CIO) or equivalent lead over all IT functions, if one does not already exist. Additionally, based on a review of individual strengths

across the existing IT organization(s), the assignment of first tier IT leaders under the CIO is an ideal interim step before moving any other staff. This approach facilitates the transfer of budgets to the centralized IT organizations and provides a small, dedicated, and qualified set of resources for the development of the processes necessary to truly integrate the existing IT organizations from a planning, development, and business analysis perspective.

3. The establishment of any program management functions and their initial enterprise processes is typically ideal before attempting to integrate applications/systems management resources and responsibilities.

4. The migration of applications development staff or control over development contractor pools to a common applications/systems management function is often one of the most disruptive activities particularly if each sub-organization to this point has had their own group of developers or contractors. To mitigate concerns in this area, a primary application lead is frequently transferred into the centralized organization for each of the major groups. These individuals can then double as the interim liaison to consolidated program management functions.

5. The introduction of functional managers should be considered carefully. It frequently takes time for new processes and organizational structures to stick. Putting in managers prior to being able to provide them the mechanisms to manage the organization could be mutually detrimental. Further, current staff may react adversely to the rapid addition of new management if not planned and timed carefully.

Another consideration is the geographic location of existing resources. Geographic diversity frequently represents both an opportunity and challenge to centralization efforts. When IT resources are dispersed throughout multiple locations, strong communication channels along with system-driven policies and processes are keys to implementing controls and enforcing standards that help control spending. However, an established IT presence in certain markets provides an opportunity for an IT organization to centralize functions in a geography that supplies the best skill sets at the most competitive rates. Remote locations may, for example, represent the lowest cost, highest quality alternative for enterprise-wide delivery of certain IT services over other locations located in higher cost-of-living areas.

Sequencing Centralization

Although centralizing IT activities can be one of the most substantial transformation efforts an organization can undertake, it is likely to deliver the largest levels of efficiency and cost reduction when implemented successfully. Organizations assessing IT centralization typically sequence the effort by:

1. Determining the complete scope of functions to be included in the centralized organizational structure and finalize the future state organizational structure itself.

2. Creating a discrete program management function and aligning existing program and project management functions under it. The existence of this function in the short term is often required to properly develop and implement consolidated project approval and planning/project management processes and methods.

3. Consolidating IT finance functions, including budget, requisition, and contract management activity, into a single technical business management function. The IT finance functions must be properly and authoritatively aligned in order to facilitate the type and detail of analysis which would be required to rationalize any investment processes and improve the procurement approach from an enterprise-wide perspective.
4. Organizing components of customer support, desktop engineering, and server operations into a single operations function.
5. Consolidating systems development activities under a single systems management function.

7

Establishing Sound IT Governance

Efficient IT organizations do not rely solely on a single individual, such as the CIO, to determine where and how the organization should invest in technology to best meet the needs of the organization. Instead, mature organizations typically leverage a group of individuals from both the organization's lines of business and the IT organization to work in concert to direct, at the highest levels, where the organization should invest for maximum return or impact. This approach, frequently referred to as IT governance or steering, is intended to control and direct IT investments to their highest value and ensure the organization's requirements are effectively met.

Most organizations possess some level of programmatic oversight related to IT investment management, but the coordination and quality of these capabilities often varies widely. In some cases, organizations treat IT governance functions simply as mechanisms to track system investment and other data, however even in these cases the systems often lack information related to the complete portfolio of IT investments or the information is considered out-of-date or inaccurate by the organization because the currency of the information is not a critical path requirement (such as getting funding).

In other organizations, broader governance functions exist, including the use of steering committees or review boards. However, the processes are often insufficient to qualify as a true control activity because they are not followed for many system investments; or they simply don't cover many non-systems related projects such as server or storage upgrades which frequently represent a substantial amount of spending and activity for an organization.

To realize efficiency from a fully-centralized governance and program management function, organizations frequently establish or enhance governance in four critical areas by implementing:

- A standard, centralized chartering process for the initiation of new programs or activities.
- A single steering committee for the review and approval of all IT related projects.
- A scheduled review process for all planned and ongoing projects as well as existing systems and functions.
- A consolidated view of project statuses including actual vs. budgeted investments by project phase.

Use of these centralized governance components often deliver numerous benefits including reduced spending stemming from a reduction in duplicative efforts. In organizations that lack mature governance for example, various IT groups are frequently engaged in similar overlapping activities, from asset management to developing the same applications. Consolidated oversight of these projects frequently reduces the level of effort needed for each of these independent activities by identifying the redundancy and eliminating or consolidating the overlapping investments.

Centralized and mature governance also frequently results in a consolidation of an organization application portfolio. In organizations lacking mature governance processes, various groups frequently develop and maintain a diverse set of custom and package applications to satisfy the same requirements. A centralized review of these disparate systems allows for the consolidation of many of these systems and their support structures as well as the retirement of systems that are under-utilized or no longer support the organization's objectives.

Sound programmatic oversight also improves ongoing project management and investment controls. Projects handled by a variety of technical and business resources in the absence of appropriate programmatic controls typically develop their own mechanisms for tracking spending and progress. These disparate approaches frequently inhibit the organization's ability to track spending on both products and resources on an organization-wide basis. Additionally, a comparison to budgeted amounts, if they exist, is rendered unfeasible in many cases.

Implementation of Governance and Programmatic Oversight

To establish sound governance over an organization's IT spending and activity, a number of components are required, including:

1. Establishing or enhancing a discrete and unified governance function, including:
    - Creating a single centralized investment council.

- Determining membership and roles within the council.
- Developing and institutionalizing IT investment control policies.
- Integrating the governance approach into established IT and acquisition processes to ensure effectiveness and compliance.

2. Implementing the required organizational support components, including:
   - Creating a discrete planning and project management group and aligning any existing program management functions under it.
   - Aligning IT finance functions (budgets, contracts and procurement) into a single organization.
3. Developing and implementing supporting processes including:
   - Developing and implementing Planning and Project Management processes.
   - Creating and enforcing a project approval and initiation process.
   - Developing review and support processes.

In most organizations establishing new governance functions, it's important to set initial performance and impact expectations low to provide time for the members of any governance groups to 'eat the learning curve' as they develop a better understanding of the IT landscape necessary for sound decision-making.

8

Standardizing System Lifecycle Processes

The acquisition, planning, implementation, support, and retirement of IT applications, referred to collectively as the system lifecycle, is often one of the largest areas of IT spending for an organization. The standardization of system lifecycle processes as an efficiency lever focuses on achieving enterprise-level delivery consistency across all phases of the lifecycle. In addition to providing a framework for efficiently managing individual systems-related work, a standard system lifecycle management approach can provide organization-wide visibility into system-related resource utilization as well as financial and project status.

A standardized system lifecycle management approach enables an organization to fully leverage knowledge, skills and resources effectively across an organization's entire IT systems investment portfolio. Standard processes also help focus efforts on key spending areas to achieve process improvements, efficiencies, and controls in the shortest timeframe. A unified, enterprise-wide management approach to the system lifecycle also helps position the IT organizations as a well-managed agent of innovation within an organization while helping leverage IT industry best practices to the maximum extent possible.

To achieve efficiencies through standardization of the system lifecycle, organizations typically focus their efforts in five major areas:

- System investment approval and initiation
- Planning and project management
- The software development lifecycle (SDLC)
- Requisition and procurement
- Ongoing system support

By standardizing each of these processes and coupling the effort with other initiatives associated with improving IT efficiency or effectiveness, organizations frequently move closer to achieving their overarching goals of improved efficiency, increased ability to leverage system and human resources, better alignment of systems with business objectives, and improved ability to manage IT as a whole.

System Investment Approval and Initiation

System investment approval and initiation processes are typically engineered to provide consistency across all IT chartering activities within an organization and help ensure that all non-operational IT support activities are formally approved before investment begins. The creation or enhancement of any existing chartering processes provides a stronger focus on project-oriented development efforts which, without appropriate chartering processes, are frequently treated and managed similarly to operational activities. System investments should be approved and funded based on explicit information contained in the charter for that project, including:

- Defined project scope, general requirements, and sponsorship.
- Estimates of required resources and effort, technical architectures, milestones, and anticipated business benefits.
- Estimates of expected ROI and/or non-quantifiable benefits as well as total cost of ownership.

In efficient organizations, charters become the basis for achieving IT and line of business buy-in and approval, and act as the central point of reference for budgeting and forecasting. Any other organizational requirements for investment funding, such as those required by general organizational leadership or an organization's finance and accounting functions, should be aligned so that existing efforts are enhanced and complemented rather than duplicated.

Planning and Project Management

Planning and project management processes are typically established or enhanced to implement rigorous controls around systems investments as well as provide increased visibility and accountability for system-related projects. A centralized planning and project management (PPM) function helps establish a set of common system development and support tools and methods used by all system efforts to oversee the project. This typically includes both the development and operational components. PPM functions are typically responsible for:

- Managing reporting and tracking of work in progress for all IT projects.
- Coordinating internal and contractor teams to manage and schedule project efforts.

- Working with any Operations Management functions to ensure an appropriate level of preparedness and support for new/revised systems and promotion of releases into production.
- Ensuring all project documentation is complete and up-to-date.
- Helping coordinate user involvement from requirements gathering through user testing to promote benefits realization.

Software Development Life Cycle (SDLC)

Given the significant and continuous investment most organizations make in custom software-based systems, a well-engineered development approach along with the ability to leverage outputs for multiple initiatives is critical to increased efficiency. Most efforts to standardize the development activities within the systems lifecycle include:

- Defining common SDLC paths and standardized environments for all development efforts.
- Leveraging a consistent project planning framework including standard templates so that project planning and resource estimation is consistent across different development groups.
- Implementing or adopting common, platform-independent tools, eliminating the cost of buying/building/maintaining multiple tools as well as making it easier for development efforts and outputs to be leveraged across multiple groups or systems.
- Developing IT coding, user interface, documentation, and terminology guidelines.

Typically, even with these components in place, each development team may evolve specific standards appropriate to their systems and technologies within the broader guidelines established. Additionally, provisions are typically made for emergency bug fixes or other situations which require immediate action, so that the processes themselves do not limit the ability of the organization to react quickly to requirements.

Requisition and Procurement

Efficient organizations typically implement streamlined, centralized requisition processes to provide visibility into organizational spending at a detailed level, as well as provide appropriate controls that ensure a standardized and efficient operating environment. To achieve significant efficiency in this area, an organization will ideally:

- Centralize and manage all technical procurement within the guidelines and policies of the broader organization's acquisition policies and processes including leveraging the mechanics of existing acquisition functions provided by the organization. The goal is to inject control and specific technical acquisition expertise rather than duplicative procurement functionality.

- Introduce a strategic sourcing discipline, seeking to leverage volume discounts wherever possible.

- Manage all hardware, software, and service vendor relationships as well as existing contracts or purchasing vehicles.

Any asset management functions within the organizations should track all existing hardware and software assets and would work closely with contract management functions to leverage the organization's purchasing power. Additionally, a single purchase requisition system should be utilized and interfaced with project tracking and budgeting systems for even greater activity and spending visibility across the system lifecycle.

Ongoing System Support

To ensure consistently high end-user satisfaction while minimizing the level of effort associated with supporting a given system, a standard process for administering end-user support should be instituted. To achieve efficiencies in this area, organizations typically:

- Clearly define responsibilities for support tiers and escalation procedures and ensure their communication throughout the organization.
- Analyze the nature of help desk activity and other support efforts to improve support and provide information for potential system enhancements or fixes.
- Make technical knowledge management a priority across systems and maintain and catalog documentation in a consistent manner.
- Prioritize opportunities to facilitate user self-service wherever practical.
- Develop Service Level Agreements (SLAs) in conjunction with the users and use SLAs as part of overall support performance reviews.
- Develop and monitor metrics as part of periodic performance reviews and recognition programs for related IT functions and vendors.
- Implement performance management across all aspects of support, e.g., root cause analysis,

continuous process improvement, trend analysis, recurring problem analysis, etc.

- Develop comprehensive change management processes which comprise various support requirements for all system implementations, including required training for staff, required documentation, and knowledge base artifacts.

Of equal importance to support processes is the creation of customer relationship processes, including service level, performance monitoring, and reasonable charge-back mechanisms. These other processes act as critical integration points helping ensure the IT organization is closely aligned with the requirements of its customers.

Other Considerations

Similar to centralization activities, standardization of systems lifecycle processes should seek to minimize disruption to the end users of any existing systems. As standardization efforts move forward, key line of business leaders who rely upon the organization's systems should be engaged early and often throughout the process to help achieve and maintain buy-in.

To a large extent, the standardization of system lifecycle processes results in a significant enhancement to a culture of accountability within and across systems-related IT functions. Efforts should involve and rely upon collaboration with the organization's IT resources to identify, coordinate, and implement any process and system changes and if required, organizational changes. System process definition or revision efforts typically require a concerted effort and dedicated time from core IT resources as well as the buy-in of key existing vendors. Each process should be reviewed, defined, and communicated prior to implementation.

Sequencing Systems Lifecycle Standardization

Organizations that implement standardized system lifecycle processes typically appoint functional leadership over enterprise-wide systems lifecycle processes prior to the development and adoption of newly engineered processes. In addition to a lack of authority and accountability without them, these individuals represent both key architects of the processes as well as the key drivers for their adoption across the enterprise.

Once leadership is established and the tools and processes in place for organization-wide adoption, it is important to implement the appropriate controls to ensure adherence is required – this requires that processes be explicit and that mechanisms exist to determine compliance, handle exceptions, and that consequence exist for lack of compliance.

Additionally, continuous improvement of the systems lifecycle processes is almost always required. Each broad process as well as specific activity processes are typically best refined over time and should evolve in both efficiency and effectiveness as they are used in practice.

9

Using Enterprise-Wide Architectures

Another major lever of IT efficiency is the use of standardized enterprise-wide architectures for a broad range of components related to an organization's IT investments. The standardization of an organization's approach to various aspects of technology typically enables significant re-use of existing capability and a material reduction in resource requirements. Regardless of industry or organization type, the fundamental components of IT systems and the components used to support them are, for the most part, identical.

For example, most of a system's core functionality is the same as any other, such as the need to store data or documents, the need to capture information via forms presented to the user in an interface, the need to search the system, create reports from the data captured, route messages based on the status of various data or documents, and so on. At their core, most systems require some kind of database, an application engine to process information, and an interface that enables user interaction with the system. To the extent that an organization can maintain consistency and standardization within these components as illustrated below, required investment on a system-by-system basis is reduced.

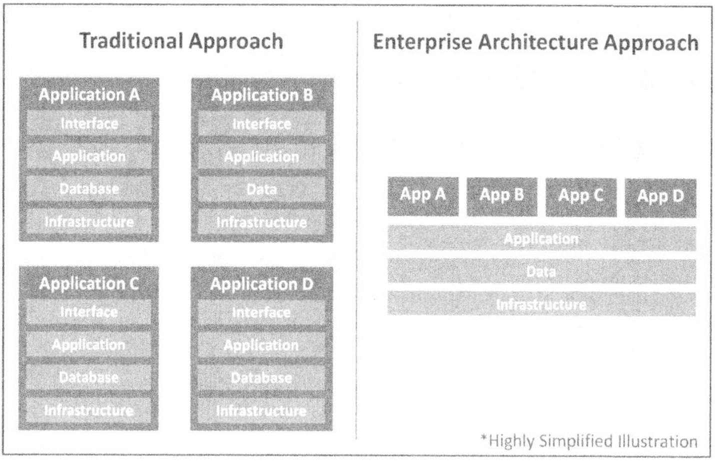

Figure 8: Enterprise architecture at its core is about maximizing reuse by leveraging a standard approach for building and managing systems and data.

The concepts of enterprise architectures also apply to the infrastructure an organization uses including its processing and storage components. Increased diversity of components, from vendors to configuration types typically translates into increased IT spending. Conversely, a single or reduced vendor and make/model environment using standard configurations typically results in a broad range of resource and asset efficiencies.

Another critical component of a standard architecture approach to IT is the way an organization integrates systems or components of systems together from a data exchange perspective. Mature IT organizations embrace the delivery of information from a system as simply a 'service' that another system can access with the right electronic credentials. This approach, referred to as a 'service oriented architecture' relies on a number of technical and information organization concepts which themselves are focused on enabling an organization to further simplify and standardize its approach to deliver IT capabilities.

The concepts of enterprise architecture are essentially applicable to almost every aspect of an organization's IT investment and operational activities. For example, an organization can develop an enterprise-wide architecture for output management, governing its approach to printing, copying, and scanning on an organization-wide basis. Based on the development of a target architecture, the organization may move towards a significant reduction in personal printers in the environment, strategic placement of network printers, reduction in printer makes and models, and the use of network technologies that enable printers to automatically notify a help desk they need repair or supplies. The result of the establishment and implementation of an output management architecture is typically lower investment in printer assets and consumables as well as increased uptime.

It's important to note that leveraging common architectures is not the same as leveraging the same application for different purposes. A system can share common computing and storage infrastructure, and/or a common database and application platform but exist as separate systems meeting very different purposes. In most situations, application and database platform standardization represents the best approach for reducing implementation risk as well as achieving desired efficiencies in development and operation. With a set of platform standards associated with content and data management, workflow, reporting, and user interfaces for example, an organization can readily leverage common infrastructure, resource skills, and authoritative information sources. Additionally, the organization maintains the ability to customize and configure these components while minimizing many of the common pitfalls associated with pure custom development, such as lengthy delivery schedules, poorly written code, limited or no external data interfaces, and shallow technical documentation.

Frequently, the best path for an organization to reduce spending on systems development is to embrace platforms upon which they can create and customize function and problem-specific capabilities. As organizations continue to move towards shared and cloud-based solutions and environments (next chapter), they should look closely at the concept of Platform as a Service (PaaS) providers as more strategic architecture components to rely on over time.

Implementation of Enterprise Architectures

Organizations seeking to embrace enterprise architectures should consider using an evolutionary approach centered on the development of target architectures and incremental implementation of technical capability as new systems are created and infrastructure investments are made. Most organizations follow a similar pattern to achieve architectural efficiencies including:

- Developing target infrastructure, application, and other architectures.

- Seeking to tactically reduce the diversity of product makes, models, and versions that address similar requirements.

- Implementing unified vendor, model, and configuration approaches to server, environment instance, and other hardware.
- Implementing tiered storage architectures relying on a number of information storage approaches to meet performance requirements.
- Implementing one or fewer common database platforms for shared system use.
- Implementing strategic systems and application platforms associated with information processing, workflow, reporting, search, and interface presentation.

Summary

This brief introduction into the concept of enterprise architectures as a spending reduction lever is not meant to represent a comprehensive technical overview of the concept, which itself encompasses multiple significant bodies of knowledge. However, embracing the concept as a general objective of an efficient IT organization and contemplating how it can be applied across every functional and technical area of an organization's IT footprint, typically reveals substantial opportunity for material cost reduction.

10

Outsourcing Systems and Service Delivery

The use of outside vendors and resources is another major IT spending lever available to organizations. When organizations use outside resources, technical capability, or capacity, they frequently realize savings in several areas. First, when using outside resources to supply certain skills, an organization may save by paying only for the level of effort of those resources required, instead of paying for one or more full-time employees. Second, an organization can take advantage of increased skill sets which can result in savings from reduced time required to complete tasks or higher quality outputs from the resources. Additionally, organizations can save by leveraging shared administrative and infrastructure resources when using services such as hosting where resources and capabilities are leveraged over multiple customer-bases.

Organizations that leverage external resources or hosting of various technical capabilities typically use one or more of the following approaches, representing increasing degrees of leverage:

- External resource expertise (contractors) to augment or replace in-house resource capability and capacity.
- External infrastructure (server and storage) including facility or infrastructure capacity for functions such as disaster recovery.
- External platform hosting whereby a vendor runs a customer-specific instance of a technology platform that an organization can rely upon to build its own system or application functionalities.
- External systems/application hosting whereby a vendor runs a customer-specific instance of an application for an organization's use.

These vendor provided-services (with the exception of simple external talent) are frequently referred to today as 'cloud-based' services. Although the 'cloud' has been around for several decades, its acceptance and near ubiquitous nature for certain services is increasing its adoption in a variety of ways by organizations. Software-as-a-Service (SaaS), Platform-as-a-Service (PaaS), and Storage-as-a-Service (STaaS) capabilities, for example, enable an organization to outsource varying degrees of requirements to take advantage of the aforementioned savings benefits.

Other Considerations

The single biggest factor dictating the decision to use external resources or technical capability is one of delivery risk. Many organizations for example have over the years attempted to outsource their entire IT organization to external providers with varying degrees of success. In cases where the transition was unsuccessful, the organization often has to go through a long and painful transition to insource the capability once again.

In other cases, organizations may outsource management and administration of only larger-scale systems such as Enterprise Resource Planning (ERP) environments which often require specialized expertise running into the millions of dollars annually. However, in all cases an organization is essentially trusting a vendor to provide a service upon which some part of it business will rely. In doing so, organization must carefully consider a number of risks including:

- **Delivery risks** - Will the organization provide the current (and promised future) capabilities, capacity, performance, and availability required?
- **Requirements risks** - Will the services meet the requirements and is there sufficient flexibility in the service to address different or changing needs?
- **Vendor stability risks** - Will this vendor and this service offering succeed and what will the organization do if it doesn't?
- **Security risks** - Is the information secure now and in the future even if, for example, the vendor is purchased or goes under?

To mitigate these risks, organizations should proceed with caution when establishing and negotiating contracts with providers. Organizations should also heavily leverage an incremental approach before attempting to leverage cloud-based services on a wholesale basis.

Summary

For many organizations, especially those with mature, centralized internal IT functions and sound IT acquisition capabilities, outsourcing capabilities and moving services to cloud-based providers represents the next frontier of efficiency. However, organizations at any level of IT spending maturity can accelerate savings by selectively leveraging external providers where appropriate.

11

Consolidating Infrastructure

In many organizations, technical infrastructure, defined generally as server, storage, and networking capability and capacity is deployed on an as-needed basis in response to additional requirements such as the deployment of services, applications or web sites. The frequent result of this organic approach to infrastructure is a data center environment comprised of a wide variety of makes, models, and configurations located in a variety of locations.

Organizations with little or no proactive infrastructure capability or capacity planning often benefit from a targeted effort to consolidate infrastructure while standardizing the platforms used to provide resources. Further, the concurrent implementation of consolidated and standardized planning and administration processes frequently adds business benefits in addition to lowered investment in equipment and support resources. Consolidation of infrastructure frequently enables an organization to:

- Minimize unnecessary investments in over capacity.

- Lower data center related support expenses.
- Optimize the utilization of server, storage, and other resources and implement better capacity planning.
- Improve the leverage of resources and skill sets.
- Reduce downtime and increase service levels through a more flexible operating environment.
- Facilitate the establishment of metrics-based service level agreements.
- Simplify the planning associated with disaster recovery and business continuity planning.
- Reduce individual location-based infrastructure requirements.

The realization of these benefits from infrastructure consolidation often results in immediate and long term savings for IT organizations.

Consolidating Server Infrastructure

The consolidation of server infrastructure can occur across a number of dimensions, with increasing levels of complexity and savings. These are typically represented by several levels of opportunity:

1. Consolidating servers that host the same applications.
2. Consolidating servers across applications that share common application components, such as databases and other software services that maintain compatible characteristics.

3. Consolidating servers that host development or test environments used for application programming or integration.

4. Migrating disparate applications to a single platform consisting of a single or reduced set of vendors and server models for the purposes of further consolidation. For example, this activity might include migrating some applications from one flavor of UNIX to another, if supported.

5. Leveraging 'virtual' environments in which applications run in separate instances of an operating system on a single server using virtualization software from a handful of vendors.

6. Consolidating data center locations. This activity might include the closure of multiple data centers and/or 'server closets' and the relocation of the infrastructure to a reduced number of locations.

Organizations sometimes face a variety of challenges in implementing the above types of consolidation. For example, an organization's level of decentralized IT control may limit its ability to accomplish even the simplest of consolidation activities. Additionally bandwidth constraints at one or more remote locations may limit the ability to locate servers off-site. However, in most cases, even level one and some level two activities result in material savings for an organization.

Consolidating Storage Infrastructure

Similar to server environments, storage capacity may also exist independently at various locations and in a variety of forms including disks internal to each server, direct attached storage units, or even multiple storage area networks (SANs). Additionally, tape storage capacity may also exist in a range of internal, direct attached, and network-attached tape libraries.

When storage infrastructure evolves organically over time, different locations or different IT teams frequently address storage needs in disparate ways with solutions acquired from different vendors supplying different technologies. The result is often a growing disparity in storage technologies that further inhibit future opportunities for consolidation and centralized management. In environments such as this, organizations often find themselves continuing to invest in new storage capacity when existing resources already possess sufficient unused capacity, but are too silo'd to be used outside their originally intended purpose.

The implementation of enterprise-wide hierarchical storage architectures, which addresses both storage infrastructure as well as policy-based data storage guidelines, can significantly reduce the ongoing costs of an organization's investment in storage and its associated support. The implementation of a new storage architecture however often requires the immediate investment in additional capacity in the form of disk and tape library infrastructure as well as training in the skills necessary to manage a consolidated storage environment.

If implemented successfully however, organizations can realize savings from reduced administration, increased capacity utilization, and the ability to shift less used information to slower, lower cost storage when a tiered architecture is used. Additionally, many organizations that migrate to a consolidated storage infrastructure approach realize additional savings from lower costs associated with data, content, and records management and administration.

Effective storage consolidation requires significant planning, including:

- Inventorying the existing storage footprint to determine the current architectures, technologies, and makes and models in production.
- Capturing current capacity and utilization levels.
- Forecasting three, five, and seven year storage requirements.
- Establishing performance and availability tiers and assigning tiers to application and data types.
- Investing in consolidated, standard storage platforms.
- Consolidating locations with adequate bandwidth and assessing ability to upgrade bandwidth, where required, for additional consolidation.

Summary

Depending on the level of investment an organization has already made in its infrastructure footprint, a phased approach that aligns platforms and standards as existing technology is retired and replaced often mitigates the level of additional investment required. Additionally, beginning with an approach that addresses a single location's opportunities while planning multi-location consolidation often facilitates broader implementation over time.

VOLUME 2 INTRODUCTION

OPERATIONAL OPPORTUNITIES

Volume 1 of Cutting IT Costs focused heavily on major, and in many cases, transformational changes an IT organization can undertake to create significant efficiencies in the way it manages investments and related operations and support functions. However, because of the scale of change required for most of these efficiency levers, the ability to take advantage of these opportunities typically lies in the hands of a select few individuals with sufficient power and control within an organization to affect the required level of changes.

The cost cutting ideas described in this volume focus instead on more operationally-oriented opportunities with accessibility to a greater number of individuals within an organization. These opportunities relate to a much larger extent on the specific operational and support approaches an organization uses to manage their IT services and assets. Although the levels of savings may be lower than larger scale changes, operational changes focused on increasing efficiency by reducing or avoiding unnecessary IT spending are still frequently material to an organization's bottom line.

Operational opportunities are also, by their very nature, easier to take advantage of, given the reduced level of change required by the organization to realize the savings. In many cases, operational opportunities are synonymous with the phrase, 'physician, heal thyself' as most of the required changes occur exclusively within the IT organization itself. However, like most opportunities described in this collection of cost cutting ideas, the best approach is frequently an incremental one based on alignment of spending and activity with a set of over-arching goals and architecture targets.

12

Improving Requisition Control

Many organizations maintain a highly fragmented approach and set of processes for requisition of vendor products and services, frequently using separate workflows and processes for approval by a disparate set of individuals. The lack of consolidated and centrally managed requisition and acquisition functions frequently inhibits an organization's ability to manage spending at the point of initiation, where significant opportunities exist to shape or even avoid purchase. Additionally, the ability to implement standards as well as gain broader visibility into organizational spending is often made unfeasible under a decentralized approach.

To address these issues, organizations often implement and use enterprise-wide standardized and enhanced requisition processes. The approach provides numerous benefits to an organization, including:

- Improved visibility into the operations of the IT function, enabling improved overall decision-making.

- Lowered product spending through improved analysis and forecasting made possible by visibility into all IT vendor spending.
- An ability to take advantage of enhanced vendor relationships through better visibility into an organization's total spending with specific vendors.
- The ability to embed product standards and other policies and business rules into the requisition system for improved standardization and control, which in turn lowers the total cost of ownership on each procured asset or service.
- The ability to better categorize requested line items for improved reporting on spending by products, functions, projects, and vendors.

The implementation of a centralized, organization-wide requisition process in the short term is a critical step in any organization's attempts to manage spending. Without the visibility and control provided by a central mechanism, it is difficult for organizations of almost any size to effectively manage spending.

When attempting to align acquisition activities within an organization with disparate approaches, taking steps that avoid significant change management issues during the initial stages of the effort frequently increases the likelihood of success. Appropriate steps include:

- Creation and enforced use of a single requisition system that captures additional key information about each line item (type, service, project, etc.)
- Approval workflow that includes organization-specific approval steps as well as system mandated

steps that require the approval of pre-defined management and technical personnel for each request.

- Implementation of a catalog requiring requestors to select from pre-approved lists of products for certain categories with exceptions handled by additional justification fields and approval steps.
- Implementation of an integrated receiving function which ties procurement details to the requisition, allows for the adjustment of key fields such as final pricing data, and captures key asset details for direct or subsequent feeding into an asset management system.
- Implementation of a reporting layer that enables management to view spending by various dimensions for improved management and decision-making.

An initial implementation that standardizes the technology and process layers, while leaving existing budgetary and organizational structures in place, often allows an organization to realize many of the benefits of consolidated procurement without significant organizational disruption. Once integrated, leadership can then use the visibility obtained to assess whether additional budget control would produce appropriate returns.

13

Improving De-Provisioning Processes

Many organizations incur monthly charges for a wide-variety of IT services based on a number of subscription or other variables, such as headcount. The provisioning of these services is typically unavoidable as new employees, contractors, interns, and other staff require access to e-mail, phone, and other IT services. Additionally, as organizations continue to rely more heavily on cloud or outsourced providers for services, the need for external provisioning of services is increasing.

The issue that arises from an IT spending standpoint is that while services are frequently unavailable to a staff member without being added to the system or list used by the vendor for billing, there's often no reverse trigger. The fact that staff require these services creates the internal 'noise' ("I need a phone!") to add the individual to the service or frequently creates an impetus for automating provisioning based on on-boarding processes. However, organizations often fail to implement straightforward mechanisms to de-provision these services when a resource leaves or they are no longer needed. Once the resource no longer needs the service, there is often simply no internal trigger, no 'noise' that tells the organization to pull the information from a variety of lists.

Given this frequent reality, organizations at a minimum may pay for services for departed or transferred individuals for months, and at worst may pay for those individual's services until a spot audit is performed that identifies the charges as unnecessary. This issue is often exacerbated by many vendors' operating structures which may require an individual to work with multiple 'service managers' or complete multiple steps to cancel different kinds of services.

The introduction of a sound de-provisioning process, along with the identification of a mechanism for identifying individuals who need services de-activated can often reduce an organization's cloud, subscription, or related types of spending. Ideally, de-provisioning processes are strongly linked to the organization's human resource management system and processes. The system(s) should initiate a number of automated processes to terminate services. However, improvement of de-provisioning often requires:

- Centralized service management within the organization, including the development and adoption of specific organization-wide policies associated with requisition and usage of provisioned services.
- A mechanism for identifying the departure of staff and, if applicable, contractors and other temporary or transient resources.
- A system and associated processes for notifying vendors and tracking services and their de-activation status by individual.

One mechanism for identifying services requiring deactivation is a lack of time charged for a specified duration in an organization's time management system. The absence of any time charged over two or more periods for example, would put the individual in a 'research' queue to determine whether de-provisioning is necessary. Additionally, a straight-forward system for tracking vendor charges, the metric upon which the charge is based, and the status of de-activation would ensure that services are terminated in as efficient a manner as possible.

Working with vendors to improve the de-provisioning process may be helpful as well. While many larger-scale vendors have taken steps to improve the provisioning process they often have not taken any substantial steps to improve the de-provisioning process based on a lack of explicit incentive to do so.

14

Unifying IT Management Methods and Support Tools

A standardized, methodological approach to the selection of IT development, management, and administration software and hardware tools is a key component in making an IT organization a differentiated asset for an organization, rather than just a cost center. This approach to the selection of tools also improves the ability of the organization to adapt to new requirements or market changes more readily, as well as leverage knowledge, tools, and resources across the organization.

Many of the broad observations in this particular area of savings were touched on in various sections within Volume 1. Specifically, the processes associated with the software development life cycle (SDLC) and requisition functions are key components in the rationalization and standardization of technology and infrastructure within an IT organization. Building on the recommendations made in those sections, the following are frequent initiatives specifically related to increasing the efficiency of managing technology:

- Implementation of a set of standards for common hardware and software needs which is embedded in requisition and SDLC strategies.
- Rationalization of disparate needs within the organization with a focus on return on investment and real business value, so that when there is a variance from standard, the impact both long and short term is well understood.
- Development of a consolidated infrastructure approach which matches compatible system uses in order to economize on cost, floor space, and maintenance without compromising service levels.
- Establishment of a common framework for business continuance and availability across applications and data centers.
- Adoption of an end-to-end functional view of systems, rather than a view of systems as a collection of point solutions.

Given these objectives, efficient IT organizations embrace and leverage a functional view of IT related technical systems and tools. A functional perspective, as illustrated in Figure 9 on the following page, highlights the necessity for a life-cycle approach to technical automation and emphasizes the need for integration between systems in order to effectively manage technology.

Not meant to be all–inclusive, this illustration attempts to focus both on the nature of technology required to effectively manage the functions of IT as well as the integrated nature of systems and applications focused on related IT functions.

Organizations that use a unified IT management and support approach across these functional areas typically find that they are maximizing the use and value of existing applications and hardware.

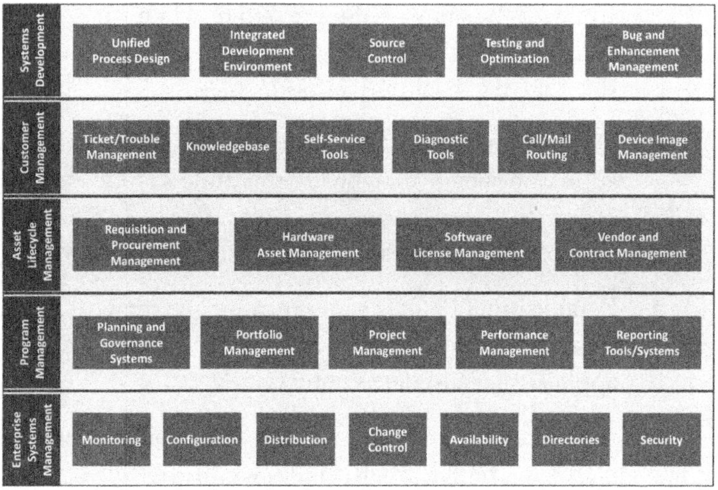

Figure 9: Standard methods and tools across a broad range of IT support and service areas saves money and improves the effectiveness of technology investments.

Additionally, the approach helps the organization more readily see systems and support as end-to-end capabilities, not point-solutions. To achieve this state, organizations typically seek to:

- Develop an infrastructure and applications consolidation mindset, benefiting the organization by maximizing current investments, better defining future investments, and reducing the overall complexity of management and administration of related technical architectures.

- Implement systems on consolidated platforms and develop them using tools and methods which are common across all efforts.
- Adopt standards for hardware and software to improve the ability to leverage physical and human resources within the organization, and to streamline and lower costs within the procurement process.
- View systems relative to end-to-end business processes rather than as point solutions for specific technical problems.
- Maximize investments in existing systems through better utilization of delivered functionality rather than additional custom development.
- Implement a common business continuity framework based upon results from a detailed business impact analysis.

Other Considerations

The primary consideration in moving towards a unified approach to IT management methods and tools is balancing the need to meet existing service levels while preparing a foundation for future technical consolidation, standardization and rationalization. This applies to both technology used for managing IT as well as technology implemented specifically to meet end user requirements. Even with this as an overall goal, it must be balanced with the realization that it is probably not practical or cost effective to re-code all non-standard applications or to port systems on non-standard platforms immediately. To maximize the value of the approach while achieving this balance, organizations typically consider:

- The cost of retaining non-standard systems versus the cost of rebuilding them.
- The cost of maintaining non-standard platforms versus the cost of porting.
- The availability and cost of resources to support standard and non-standard systems/applications.
- The importance or relevance of a particular technology to the strategic needs of the business.

Initial Steps

To move forward, organizations frequently initiate the effort by:

- Reviewing their current physical infrastructure for rationalization and consolidation opportunities and implementing common support tools and methods as the environment is rationalized and/or consolidated.
- Reviewing existing systems and applications to maximize and leverage the potential they already possess.
- Evolving enterprise standards for desktop and server hardware and software support, addressing the most common opportunities to unify the toolsets and methods on an incremental basis.

15

Improving Hardware and Software Asset Management

A mature asset management function is another important component of an organization's ability to effectively control IT costs. In addition to helping the organization better leverage idle or under-utilized assets, the existence of strong asset management can significantly reduce technical support and administrative costs.

IT assets commonly segment into two major categories; hardware and software. These assets serve vastly different purposes and to a large extent, exist as wholly distinct concepts; one physical and the other largely virtual. Based on the unique purposes and nature of each, organizations typically realize increased efficiency by managing each using different approaches and even systems.

Hardware

Organizations frequently maintain some level of hardware asset management capability. However, this capability can vary widely, with some organizations simply using manually maintained spreadsheets or occasional outputs from network discovery or management tools. In some cases, even within very large organizations, a centralized inventory of physical technical assets simply does not exist.

Even when asset inventories do exist, they are frequently not inclusive of all hardware assets and in many cases, do not track many valuable details such as asset responsibility, location information, or purpose information. Another common issue is that no individual or role is often specifically tasked with the responsibility of maintaining a complete and accurate asset inventory, making the repository a virtual 'tragedy of the commons' in terms of data quality and accuracy.

The implementation of an enterprise hardware asset management system and associated support function frequently lowers the total cost of ownership of hardware assets while improving their alignment with the organization's strategic objectives. Spending reductions are frequently realized through a variety of savings levers including cost avoidance, improved asset utilization, reduced theft, and lower support costs. The full benefits of IT asset management are typically realized by addressing the function as a core component of the complete asset lifecycle. Critical steps for effective hardware asset management include:

- Allocating appropriate resources and responsibilities to asset management functions.
- Determining which assets will be tracked, such as the required value level, so the organization does not waste significant money and energy tracking mice or other low value assets.

- Implementing asset planning activities including requirements forecasting for larger ticket hardware purchases.
- Integrating asset management with acquisition and infrastructure change activities in order to maintain the accuracy of asset information.
- Implementing a repository with full reporting capabilities to facilitate a wide-variety of planning and support functions.

As with any inventory-oriented system, addressing the automatic capture of incoming assets and documenting changes is often of greater initial importance than documenting currently deployed assets. Without appropriate receiving and change controls, the asset inventory will represent only a snapshot of the current environment which then becomes out-of-date quickly as additional assets are acquired and existing assets are changed.

Software

Many organizations do not adequately manage software or use software license management (SLM) solutions to track software licensing. Even those that do use SLM solutions, have traditionally focused on compliance over cost savings. The implementation of an SLM system typically reduces software licensing fees in a number of ways, including:

- The recapture of unused licenses installed on systems and their redeployment to other systems in lieu of new license spending.
- The cancellation of software maintenance, assurance, and technical support on unused and unneeded licenses.

- Improved pricing through consolidation of demand for new products and version upgrades.
- Lowered risk of penalties and fines from unlicensed use of software as well as less costly compliance reporting.

In addition to reduced software related costs, the improved visibility into an organization's software inventory enables further standardization opportunities that could increase volume discounts, while also lowering support costs associated with maintaining a more diverse software product base. However, it is not atypical for SLM system implementations to be measured in degrees of failure. The most common reasons for a lack of success are:

- A failure to formally assign software asset management to an individual, instead treating it as a technology rather than a function.
- A lack of effort to adequately configure the software or its network agent(s) to recognize the full suite of products installed in the environment.
- A lack of effort to collect entitlements (purchased license information) and enter them into the system and/or a failure to keep entitlement information up to date.
- A lack of effort to reconcile the entitlements against the identified installations.

To avoid or mitigate the factors that inhibit successful software asset management, organizations typically use an approach that combines the technical implementation with formal organizational assignment of responsibility over software asset management. To take full advantage of this opportunity, organizations often undertake a number of steps including:

- Development of organization-wide requirements for license management.
- Evaluation of any existing internal software license management tools as well as an assessment of the SLM vendor landscape.
- Conducting a pilot of a small number of products in a test environment to validate solution feasibility and value.
- The broader implementation of a license management solution enterprise-wide.
- Integration of the receiving processes and user delivery confirmation with the SLM solution.
- The reconciliation of existing entitlements and software installs.
- The removal of software installed but not used.
- The cancellation of software maintenance and technical support on unneeded licenses.

The implementation of an SLM function that tracks licensing availability and use, automates deployment and installation, and integrates with other IT service management systems and capabilities represents one of the most straightforward ways organizations can reduce IT spending while increasing user productivity and satisfaction.

16

Leveraging Virtualization

In the decades since the advent of the 'Wintel' system and in some cases earlier, organizations relied upon the concept of a single instance operating system and associated system environment per server. With the commercialization of virtualization software and its maturity into an enterprise-level solution, organizations can now run multiple instances of an operating environment on a single machine enabling an IT organization to greatly increase the processing utilization of its server investments. For organizations with large server and associated datacenter footprints, virtualization frequently results in a wide range of savings, from reduced facility and energy costs to substantial reductions in hardware and support investments.

Despite its increased acceptance across organizations as a cost-effective way to deploy systems and related operating environments, many organizations fail to leverage virtualization because they perceive a risk associated with running multiple instances of an operating environment on a single-server. However, the all or nothing virtualization approach leaves considerable opportunity for savings unrealized.

For many organizations, using virtual server environments for development and testing not only results in material hardware and support savings, but represents a significantly more efficient way for developers and integrators to work. In the case of testing for example, a team can 'reset' an environment after a failed installation or test scenario with a single command, in lieu of restoring the environment using significant scripting and manual efforts previously required.

By embracing virtual environments for development and testing as an initial step towards introducing virtualization, many of the perceived risks typically associated with running production systems in a virtualized environment are eliminated. Instead, the technology is applied to the portion of the system lifecycle where the technology's accepted risk profile fits the organizational need.

17

Using Open Source Software

Open source software, which includes a wide range of software tools, utilities, and platforms available to organizations, represents a class of software that is largely developed by open communities and usually distributed free of charge for commercial use. Organizations that leverage open source software across the enterprise often realize a number of efficiencies including low or no initial and upgrade license acquisition costs as well as reduced support requirements in a number of areas. Additionally, open source software typically gives an organization just that - open source code from which they can modify or enhance to meet specific organizational requirements, rather than wait for vendors to address requirements in future versions of their software.

The debate over use of open source software is an obvious one. Many software companies have invested millions or even hundreds of millions of dollars in the development of software packages over time for purchase by organizations and individuals. Their contention is that this level of investment has resulted in superior products as well as superior support that ultimately results in a lower total cost of ownership relative to open source solutions. Many of the hotly debated areas associated with open source software adoption, both for and against, include its:

- Reliability and stability.
- Performance characteristics.
- Security vulnerabilities, or lack thereof.
- Support options, both immediate and longer term.

An interesting reality is that many organizations have taken a position not to use open source software for many enterprise-level purposes, failing to recognize that they already have. IT organizations may debate, for example, the feasibility of leveraging open source platforms for web content management while failing to realize their websites already run on open source web servers like Apache™.

To realize potential savings from open source software, many organizations look for opportunities across a broad range of areas including tools for internal IT management, simple utilities, specific applications like productivity software, enterprise platforms, and underlying technical components that become a part of broader IT solutions.

18

Improving IT Contract Management

Many IT organizations leverage a variety of external support contracts to provide IT services to their organizations. These contracts, which range from high-volume multi-supplier general service vehicles to smaller specialized shorter-duration contracts, satisfy for many IT organizations the vast majority of IT services required. Over the past decade, driven by a number of events focused on increasing efficiencies associated with delivery and oversight costs, organizations have increasingly consolidated IT contracts associated with development, data center, and seat management services into one or a small handful of very large single-award, multi-year outsourcing contracts.

While in some cases, the cost of delivering the services covered under the contract has decreased commensurate with achieving economies of scale, many other issues can arise, from reduced performance to an explosion of costs and risks in other areas. Often times, under larger-scale IT service delivery contracts, organizations face a number of new or increased risks and issues, including:

- Significantly increased contract management costs as the number and types of underlying task orders increase over time, along with ever increasing sub-contract activity.
- Reduced flexibility of service levels from 'one-size-fits-all' decisions as contract management resources, and sometimes IT leadership itself, mistake the 'good' costs of offering classes of service with the 'bad' costs of overpaying for infinite flexibility.
- Increased heterogeneity and segmentation of the IT environment driven by the decentralization of decision-making across dozens and possibly hundreds of individual task orders and their managers.
- Reduced visibility into IT spending across various categories caused by significant segmentation of activity across tasks.
- Increased duplication of efforts and redundant systems driven by a lack of centralized oversight or meaningful adherence to architecture standards.
- Increased overall IT service delivery costs as the organization trades reduced rates for increased volume caused by the need to manage an increasingly diverse and segmented approach to delivering IT services.

The concept of outsourcing what should be commoditized IT support services to one or a handful of vendors is a valid strategy. The belief though, that a single vendor by definition equals a single approach and unified set of standards, and will act as a single point of accountability, simply hasn't proven true in practice. Whether tasks are divided by sub-organization, location, system, or project, a vendor will almost always default at a minimum to whatever individual task leadership desires. This model ties virtually all performance and quality management to the capabilities of each individual task order manager. In addition, in a model that is built on an assumption of vendor internal coordination, vendors are often left with a set of alternate incentives that drive revenue-oriented behaviors that run counter to the client's larger efficiency objectives.

Traditional Approaches

Given the realities of the systemic issues associated with large-scale IT support contracts, IT leaders have reacted in a number of ways in an attempt to take advantage of the opportunities these contracts present while mitigating their risks. With an understanding of the general environment in which these contracts operate, IT leaders have applied a variety of management and oversight mechanisms over time, including:

Consolidating existing tasks into a small handful of larger monolithic task orders. In an attempt to enforce standardization, reduce costs, and eliminate redundancy, leadership may attempt to 'un-ring the bell' of allowing autonomous, decentralized tasks to flourish. While the objective may be sound, in practice the effort often results in only a contractual re-arrangement with minimal impact on the underlying direction, quality, and performance of the actual work performed.

Using Independent Verification and Validation (IV&V) or similar types of third party review. Many IT organizations use independent assessment activities to help monitor and manage larger-scale systems development or integration efforts. This approach, when performed by resources with the appropriate expertise, often enables better project course correction but the use of IV&V is frequently limited to software efforts and often times, the monitoring resources only look at progress or earned value metrics, failing to recognize signs that the project is actually 'successfully marching off a cliff' from a business benefit perspective.

Using satisfaction surveys. Another method used to help oversee performance is the use of brief satisfaction surveys typically conducted at the end of user-oriented help desk support efforts. While providing some insight into performance, their limited use does not typically generate a material impact on the overall performance of the contract.

Leveraging cost plus award fee contract structures. Many larger-scale IT support contracts are structured as cost-reimbursement contracts which include a periodic award fee structure designed to create an incentive for performance. While representing a viable strategy for incentivizing delivery performance, cost plus award fee contracts often fail to motivate efficiencies on the part of the contractor and based on historic results, typically result simply in 'institutionalizing mediocrity' across the contract's service categories.

Relying heavily or exclusively on a requirements-driven approach. The concept of relying heavily or even exclusively on customer requirements as a means for identifying objectives, solutions, and services represents another approach which has typically fallen short in practice. In most cases, the approach effectively transfers risk from those who should know technology to those who know the least about technology. In the case of software development, requirements are often defined by 'analysts' who specialize in capturing and organizing information from users instead of truly translating business needs into functional and ultimately technical requirements. The approach itself often ignores the concept of the IT organization's role as a forward-focused support arm that should drive value in part by understanding the art of the possible. This perspective is embodied by Henry Ford's famous quote that if he had asked customers what they wanted they would have said faster horses.

Each of these mechanisms, to some degree, plays a role in the structure and management of sound IT contracts. The underlying issue with each of these however, is that they are simply not holistic, and often performed by business-oriented resources that lack the necessary technical acumen to make optimal decisions.

Additionally, many organizations operate as if the existing CIO organization is structured to support and drive the very contracts upon which so much IT support is delivered. In reality, the broader CIO functions often fail to touch in meaningful ways the outputs and outcomes from these larger vehicles. The disconnect, whether caused by legacy contract management structures or other issues, often leaves an organization vulnerable to increased costs across their IT investment areas as well as reduced benefit from the IT services and solutions themselves.

Organizations using large contract structures with substantial underlying task orders frequently fail to recognize that the model itself is fundamentally flawed. The result of this approach is essentially the outsourcing of almost all aspects of IT management. The real levers of performance including governance, architecture, technical business management, customer collaboration, and others are distributed and left to disparate individuals who often do not work within the IT organization and frequently specialize in contract management over IT management – this includes the provider side where 'program managers' are also often contract or project managers with minimal operational experience in IT solution and service delivery management.

Enhancing Contract Management Approaches

To realize the intended benefits of these contracts and drive the desired impact and performance, IT organizations often establish a contract-specific oversight structure, as illustrated below, that embodies the key functions that exist at a larger scale within the CIO-led IT organization.

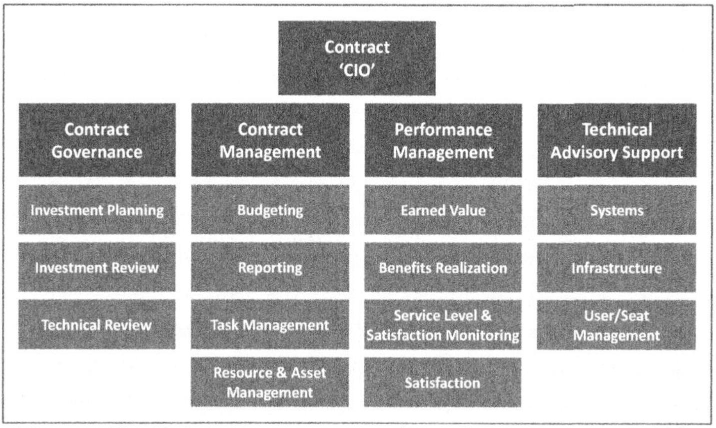

Figure 10: Driving governance through a contract-specific CIO-like function as well as appropriate oversight functions can improve contract performance while reducing spending.

Whether leveraged internally or itself outsourced to some degree (to a different provider), organizations typically require a contract oversight structure that:

- Maintains a simple governance capability which coordinates reviews, provides feedback on, and escalates, where appropriate, desired contract-specific investments and projects to the organization's broader IT governance functions.
- Possesses centralized contract and task management including frequent activity and status reporting to a variety of stakeholders.
- Acts as an integration point between the IT organization's objectives, policies, standards, and processes and the support provider's approach to service delivery by ensuring tasks are structured appropriately.
- Acts as the primary mechanism for capturing and reporting contract performance.
- Provides independent business and technical expertise in order to provide advice to the IT organization as well as the contractor to improve solutions and services as well as drive innovation.

The formality and structure of the approach frequently represents an ideal hybrid between a traditional program or project management office (PMO) and the unique operational, business, and technical structure of an IT organization. These functions, depending on the size and scale of the contract, may be performed by one, or potentially dozens of resources to fully meet the oversight requirements.

Critical Success Factors

An improved structure and approach is only as good as the resources supporting it. Successful oversight requires the individuals and supporting organizations to possess and reflect a number of critical attributes, including:

Strong technical acumen. Access to resources with significant technical and IT operations expertise, whether related to user provisioning, seat management, data center operations, systems design and development, structuring relational databases, web content management, configuring fibre-channel switches, or any other technical or operational area. The ability to recognize and articulate how support and solutions should look or can look represents a critical missing component of many current approaches to IT contract management oversight.

Contract oversight leadership with a 'CIO mentality' instead of a contract management mentality. An individual who can effectively balance a passion for leveraging technology as a business enabler with the desire to provide services in an efficient, fiscally-sustainable way.

A passion for building and running exceptional delivery models. A mentality that reflects a desire to 'wow' customers by truly understanding their business needs and identifying innovative ways to structure solutions and services that best meet those needs.

A collaborative, appreciative support approach. Understanding and embracing the oversight role as a guide, coach, and enabler in order to drive performance by working with a vendor or set of vendors productively over the life of the contract.

An ability to organize and execute technical and management activities effectively. The operational and administrative management expertise to successfully handle the daily activities associated with ongoing contract and task management.

Summary

An oversight organization which possesses the business and technical capability to understand and drive performance from an IT contract provides an organization with the resources it needs to objectively analyze proposed and ongoing solution implementations, levels of effort, proposed resource qualifications, statements of work, service levels, and other important dimensions of contract delivery.

When accomplished successfully, this oversight support frequently translates into material improvements in the value of implemented solutions and significant increases in user satisfaction. In addition, the levels of investment required to achieve this value decrease through the implementation of efficiencies that drive down cost without sacrificing quality and service levels.

Each of these benefits ultimately results in a reduction in an organization's business and operational risk while directly improving its ability to leverage technology to accomplish its mission.

19

Delaying Upgrades

Retaining desktop systems and other hardware for an additional one to three years before replacement can enable an organization to temporarily redirect investment into other areas. An ancillary benefit to delaying desktop replacement may also be the potential alignment of a refresh with the deployment of new versions of an operating system or at a minimum, the deployment of PCs that possess the technical requirements to run a future version.

A common risk in the short term with this strategy is that the speed or capabilities of the current desktop systems may be insufficient to meet user requirements during the delay. This however often proves to be an unrealized risk in practice and exception processes to deal with specific needs are frequently sufficient to mitigate the impact. Another, more profound risk is that organizations may be unable to direct funds *back* to the eventual replacement from the areas where the funds were re-directed. In many organizations, there is no such thing as temporary or a 'band-aid'. If it works, it's permanent.

For many organizations however, the risks are worth the ability to use funds elsewhere or to offset other cuts. To assess the savings potential of this opportunity, organizations typically look at:

- The productivity or performance impact and expected levels of user dissatisfaction during the delay.
- The resource requirements to perform a potentially larger upgrade of an organization's desktop fleet in lieu of an earlier initiated phased approach.
- Whether the current desktops can be used without upgrades for an additional period of time by forecasting performance, disk space availability, and the impact of application or other technology deployments on desktop or server and storage requirements.
- The estimated costs of repairs and downtime from out-of-warranty support.

Many organizations use a '3/33' or '4/25' replacement schedule where one third or one fourth of their desktop fleet is replaced each year. This strategy provides a consistent, predictable expense stream and replacement workload. From this perspective, projected savings are viewed as a function of forecasted replacement vs. actual replacement; with organizations leveraging their assets longer thus spending less on a per year basis.

20

Using a Multi-Tier Thin Client Desktop Strategy

When many organizations refresh their desktop systems, they often use what is essentially a single tier strategy, deploying the same desktop system to all individuals. The resulting homogenous environment often improves the efficiency of support and enables improved discounts based on volume. The single-model strategy is frequently unfeasible in many environments however, where there are large populations of 'standard' and 'power user' populations. In these instances, a multi-tier strategy is often used to address the diverse needs of users.

By continuing the segmentation approach and formalizing its strategy, many organizations take the opportunity to further reduce acquisition and support costs by defining and using additional desktop tiers. For illustration, three potential tiers could be:

- Tier 1 – Thin Client: A network booted minimally configured PC that is used by users with only basic computing requirements. In addition to lower acquisition costs, all administration occurs at the server level resulting in reduced support costs.

- Tier 2 – Basic Desktop: A fully functional PC with a basic configuration. This system tier is leveraged by those requiring more power than the thin client tier or the need for non-standard software.

- Tier 3 – Advanced Workstation: A more powerful PC that is leveraged by those requiring capability greater than the basic desktop.

A similar approach is also frequently used to address laptop requirements or even software, where open source or cloud-based productivity software such as word processing may be a feasible alternative. In some cases, simply piloting the thin client or open source approach can be used as a competitive lever to drive down the pricing of systems and productivity software.

To assess the effectiveness of a multi-tiered desktop strategy, organizations typically initiate a pilot to determine the feasibility of deploying different tiers within its user community. Initial pilot selection is often based on identification of a small number of administrative staff with straightforward computing requirements in a single location.

21

Using Web Content Management Software

Organizations typically realize significant benefits from using a web content management platform in lieu of the traditional webmaster-driven single file per page approach. The advantages are frequently so substantial to an organization that it represents one of the key levers an IT organization can pull that almost always increases the effectiveness of the function and reduces its cost.

Traditional approaches to managing web content frequently rely on a cadre of web editors, typically residing in the IT organization, to individually edit web pages and the styles these pages rely upon for look and feel. The approach, based on legacy technical approaches for delivering web content, often possesses significant drawbacks for an organization. First, because the web editors are required for additions or changes, they in essence represent a bottleneck which can materially reduce how dynamic a site's content is. Because most organization's web presence is now a strategic enabler, anything that effectively inhibits the flow of content to various audiences inside and outside an organization is a potentially serious weakness.

Additionally, the traditional approach is expensive. The manual, labor-intensive nature of editing web pages to update content typically requires more dedicated resources. Web editors and related IT resources represent in essence a duplicative step as they take content created and edited by others across an organization and add it to new or existing files. Additionally, organizations spend additional funds to continuously fix issues such as broken links, unify branding on different pages, and piece together analytical reporting required by the organization's management.

Web content management systems enable an organization to automate deployment of content by enabling authorized users across the organization to maintain web content directly. Based on this approach, an organization can reduce its footprint of webmasters in the IT organization or direct the resources to higher value information management activities. Web content management systems also deliver automated workflows for content approval, lock-down look and feel aspects for centralized control of branding, automatically disable broken links, and even provide user-friendly modules for adding forms, calendars, and image rotators to pages for example. Web content management systems also rely on a centralized database rather than individual page files frequently resulting in reduced expense associated with infrastructure support.

Evaluating the Value and Feasibility of Web Content Management

Organizations that have not yet implemented a web content management system to reduce the costs associated with maintaining its web presence often look at five key areas to determine fit and value of various alternatives in the market:

1. Who and how many users will need to edit content in the organization?
2. Does the organization have unique requirements, including the need to support localized or regional content, perhaps in different languages?
3. What level of sophistication does the organization require in terms of features and functions?
4. How complex is the implementation within the organization given the expected user community, existing content, and new processes required?
5. How much effort will it take to convert the existing content?

Some organizations maintain hundreds of thousands of web pages over thousands of domains, spending tens of millions of dollars to maintain each individual page and their associated cascading styles. Successful implementation of a web content management system frequently results in major reductions to this support footprint while increasing the currency and value of the information presented even in environments with extremely complex requirements.

Implementing a Web Content Management System

Many organizations find the implementation of a web content management system to be more straightforward than other system implementations based on the data already available about their web content and approaches, the mature nature of the packages (both commercial and open source) available, and the tools and resources provided by many vendors including semi-automated content ingest of traditional page files into a web content management system. Organizations that undertake the migration typically:

- Develop a high level concept of operations and future state site architecture.
- Inventory and map existing content and develop branding/page templates.
- Implement the infrastructure and platform software.
- Establish workflows in software and initiate content migration.
- Review content and site for functionality and migration success.
- Conduct training and knowledge transfer activities for successful operation and support.

Summary

In many organizations, the inhibitors to leveraging web content management systems have nothing to do with cost savings or the strength of the business case. For some, the issues have more to do with resistance from the webmaster community who fear job security, but are often the ones tasked with evaluating the feasibility of web content management systems. Organizations are also sometimes frozen from action based on a variety of opinions about which vendor or platform to select, the viability of open source alternatives, or even confusion between web and other content management systems, such as those focused on document management or content collaboration and file sharing.

However, organizations that can avoid or move past these potential issues typically find that the implementation of web content management systems represents a straightforward way to reduce IT spending while improving service delivery in an increasingly critical area of the enterprise.

22

Consolidating and Standardizing the Printer Fleet

Due to unmanaged growth in print resources stemming from few policy or acquisition controls, many organization's printer fleets consist of a diverse set of dedicated and networked printers representing a wide variety of makes and models. The result is unnecessary printing related costs stemming from:

- An inability to leverage volume pricing discounts in printers, supplies, and vendor support.
- A need to stock a larger variety and quantities of consumables and related printing supplies.
- Increased costs for support and maintenance due to a lack of standards as well as an inability to leverage a common network printer management framework.

An adherence to standard printer models across an organization, a reduction in the number of dedicated printers, and a bulk refresh of aging printers often helps reduce an organization's often considerable printer related spending. Additionally, this opportunity may extend to other shared technical resources including scanners and fax machines, especially where fax consolidation could lead to additional savings from reduced phone line related expenses. To take advantage of printer fleet consolidation, organizations typically consider:

- Creating a small number of printer classes based on requirements - Classes for dedicated, networked workgroup, networked volume, and color printing are the most common types of classes used.
- Selecting a single model for each printer class - In addition to quality, reliability, and price, model selection based on the printer's current lifecycle stage may allow an organization to purchase the same model throughout its consolidation and standardization efforts while enabling better remote management by allowing a single network management interface to be used across all printers.
- Implementing policies directed at reducing the number of dedicated printers in an organization's environment – A reduction in dedicated printers often results in a direct reduction in spending related to printing.
- Integrating a printer management system or console – Most printer management systems are capable of integrating directly with call center applications so that printer errors result in an automatic trouble ticket submission, which significantly reduces the need for manual intervention and downtime.

Additionally, organizations may wish to assess the cost effectiveness of outsourced print services offered by a variety of organizations whereby the organization pays a cost per page with print assets, consumables, and maintenance all provided by the vendor.

23

Implementing a 'Sparing Tier' for IT Support

Many organizations do not possess a policy and capability for maintaining spare IT hardware assets as an internal support tier, instead relying on vendor support to achieve the service levels required for any given hardware asset. However, the cost of support agreements that provide parts replacement is in many cases not competitive with an effective sparing policy. For example, in many cases the difference between the annual cost of basic and premium support levels is the provision of on-site spares, yet the price difference between the levels is more than fifty percent of the equipment's original cost.

By reducing vendor support levels or in some cases canceling support and using spares to affect repairs, organizations often lower their overall support-related spending. To effectively leverage sparing, organizations typically:

- Evaluate the availability requirements of common technical assets.

- Assess the ability to treat the asset as a 'field replaceable unit' as well as the ability to affect repairs without immediate vendor support.
- Determine the economic differential between levels of vendor service required to meet availability and level required if spares are used.
- Compare the differential to the additional cost of acquiring and storing spares.

At certain quantities and maintenance frequencies, some vendors may provide spares upon request at no charge to facilitate their resources' time and improve customer satisfaction. Additionally, a critical component of the effective use of spares is the use of a hardware asset management system to track and manage the additional inventory created by hardware spares.

VOLUME 3 INTRODUCTION

TACTICAL ACTIVITIES

Volume 1 of Cutting IT Costs focused on making larger-scale changes to realize significant efficiencies in the management and delivery of IT services to an organization. Volume 2 focused heavily on smaller yet still material changes to an IT organization to extract additional efficiencies. Volume 3, Tactical Activities represents a collection of highly tactical IT cost reduction activities that can be implemented with minimal effort by a wide range of individuals inside an IT organization.

In many organizations, acquisition of IT hardware, software, and services is performed by a general procurement function or in some cases by business managers or even users themselves. However, the procurement of IT products and services is part art and science requiring a combination of technical and business acumen in order to maximize the purchase opportunity while minimizing investment. For any organization regularly acquiring IT products and services, significant savings are frequently possible if the right technical acquisition skills are applied accordingly.

Take for example some common scenarios. An IT manager, developer, or user needs technology or related services from one or more vendors. In many organizations, these individuals obtain several quotes from different sources and pass the information on to a procurement shop for actual purchase. The procurement shop may use one of these quotes or buy from another source with better total pricing. A job well done. However, what if the purchase was never necessary in the first place? Or other, less expensive products or providers serve the exact same purpose? Or the purchase can be leveraged in other ways to help offset other needs? Applied technical procurement expertise often pays for itself multiple times over for any organization buying several million or more of IT products and services each year.

As in other volumes of Cutting IT Costs, the ideas described throughout this text are just that - ideas. They are not meant to represent the complete universe of cost saving techniques but instead illustrate a number of opportunities to tactically reduce IT spending in a number of areas.

24

Software Discount Programs: Are You at the Right Level?

Many major software vendors have established formal discount programs with pricing tiers based on quarterly or annual purchase volume. There are several important factors to be aware of when buying from vendors who use these programs.

First, vendor systems for tracking what a client purchases are often very poor at doing just that. As the basis for determining what tier an organization falls in, vendors frequently fail to account for multiple buying sites or groups within an organization, do not aggregate purchases from different areas of their own organization, or simply fail to count straight up sales made to an organization. These factors frequently contribute to a lower tier and lower subsequent discount.

Second, many of these programs only dictate the discount they provide to the VAR or reseller channels an organization is buying through. Hence, there is still considerable room for negotiation associated with the margin that a particular supplier will add.

Third, many of these programs have no formal system for updating an organization's discount tier based on their continuing purchases. In many cases, vendors rely on their direct or territory representative assigned to an account to update an organization's status in their system. This manual process often results in a significant lag between when an organization actually qualifies for a better discount, and when the vendor begins to apply it. Often, an organization's discount tier remains what it was initially set at until the organization specifically asks for it to be updated.

Lastly, most vendors will override the discount program to win business. Competitive situations (which every purchase should be), bigger dollar size, or end of quarter timing create a deal environment that results in significant discounts independent of the vendor's formal programs.

When working with IT vendors that use and attempt to enforce a formal discount structure, organizations should use the following rules:

Know the Program Terms. Find out what the tiers are, what volumes are required, and what the policy is for moving from one tier to the next. For example, a discount may apply for an entire quarter based on last quarter's volume, or an organization may qualify for a better discount when it reaches a certain number of 'points'.

Ask the Vendor for a Report of Qualifying Purchases. Compare that report to any available enterprise-wide purchasing detail to determine the accuracy of the vendor's report. If there is a discrepancy that would dictate a discount change, calculate when that should have occurred and request a refund or future credit for the difference.

Do Not Use Program Discounts for Anything Other than Small, Periodic Purchases. Demand vendors provide deal-based discounts that reflect the competitive circumstances of each purchase. If you do initiate a purchase where the channel price has been dictated by a formal discount program and are buying through a channel partner, ensure the best possible price by limiting the channel markup through negotiation of your final price with that partner.

Summary

Given the inability of vendors to manage their own programs, it is important for buyers to understand and manage for the vendors how these formal discount programs affect their pricing.

25

When Vendors Won't Budge on Service Costs

There's no question about it, vendors love to bundle services and support with product costs. The reason: the margins on services are the highest for vendors, and no vendor wants to make it easy to negotiate these costs. But for the purchasing organization, these services are usually the best to negotiate down exactly because of their high margins. Ensuring the vendor breaks out the pricing is obvious, but sometimes the vendor won't budge on support, training, or implementation costs. The reasons for this are numerous.

First, an organization may have been so successful at product price negotiation, the vendor's only deal profit is in the services. Vendors will often look at deal margin across the entire potential sale and too much of a discount in one area naturally creates resistance for aggressive discounting in other areas. If an organization has secured what appear to be 'too good to be true' product discount levels, the reason is often baked in service or support margins.

Second, the vendor only has flexibility if their personnel are performing the services. Often, a vendor will subcontract the work to another firm, making those costs essentially inflexible for the vendor. That inflexibility is passed on to the purchasing organization. Many vendors for example use third parties to provide training so their ability to cut pricing by very much in this area is limited. Conversely, implementation is often an area where vendors often use their own personnel and therefore may be more flexible, based on their ability to spread the associated personnel costs over multiple clients and implementations.

Lastly, a vendor may fear erosion to its support pricing or overall deal values as future prospects begin to recognize how the vendor is pricing its services and what its price flexibility is. This is a particular problem when the products the services are associated with are essentially IT commodities.

However, even when a vendor won't drop pricing further, an organization can use scope to make the money they do spend go further. Try asking a vendor who is inflexible on price if they are willing to provide additional services such as product or solution optimization, tuning, testing, or quality assurance. Additionally, try asking for project management services to reduce risk or asking the vendor to take responsibility for installing and configuring more parts of the solution stack such as databases or storage. Another area where the vendor may be willing to provide services at no charge to win a deal is disaster recovery planning including help on how to configure the product or solution for disaster recovery or even providing recovery plan detail for the system or solution.

Summary

A vendor may be unwilling to negotiate on services pricing. When this occurs, negotiate additional services or support to maximize the money spent.

26

Avoiding Sales Tax on Software

Many states now treat software as intangible rather than tangible property which in most cases means a state's sales tax laws do not require sales tax to be paid on software when the software is either electronically downloaded or transferred by e-mail and no physical property such as disks or manuals are provided. Additionally, software is also non-taxable when the software is installed on-site and the installer leaves no physical discs or manuals behind (commonly referred to as 'load and leave').

Organizations should check their states sales tax laws or consult a tax attorney to determine whether they can benefit from its state's tax laws. If so, remember when purchasing software to ensure no physical property in the form of media or documents is transferred and the purchase order is worded accordingly to avoid paying sales tax. A media content company in California, for example, was able to save $121,000 in sales tax on one large software order alone by structuring the wording of its agreement to explicitly state that no tangible property was being purchased.

Summary

Check to see if your state treats software as intangible property and whether intangible property is tax-exempt. If so, structure software purchases to avoid sales tax.

27

Finding and Fixing Invoice Errors

We've all heard the expression that the cobbler's son is the last to get his shoes. That saying has never applied more than to technology vendors' billing systems. The invoice errors these companies make can be very costly - unless you stop it. Analysts differ on the percentage of vendor invoices that contain mistakes, from 10-30% depending on who you ask, but regardless of the average these small (and large) invoice errors can add up fast.

Common Mistakes

Common mistakes vendors make on their invoices include:

Quantity errors. Vendors frequently bill and ship (or simply bill for) greater quantities than what was ordered. If the extra product actually shipped and you want to keep it, make sure to negotiate an additional discount on the overage in exchange for not sending it back. If you do not want to keep the extra product, make sure the vendor arranges and pays for shipping the product back. In many cases, the process of arranging pick up and return is often enough for the vendor to purposely or through its own negligence abandon the extra product at your site. If this occurs, you can utilize or resell the product after a reasonable time period, about 60 days. Make sure to deduct the amount of the error (short pay the invoice) or subtract the additional discount you negotiated from the invoice before paying it.

Extra product line items. Vendors will often add additional line items to an invoice, again sometimes actually shipping the product. Treat these errors in a similar fashion to quantity errors. If the additional line items are required for your product or solution to properly function, attempt to negotiate a substantial discount or no charge for the additional product leveraging the potential return of everything based on the 'newly discovered costs'.

Extra service or training line items. Vendors will also occasionally add to invoices 'integration', 'install', 'training', and other phantom services never ordered. These unordered services are also often never performed unless your organization initiates the service. Never pay for these services unless ordered. If the vendor insists the services are required, treat these line items as you would required but unordered product line items as described above.

Shipping charge errors. Shipping charge errors typically fall into two categories; a shipping charge when nothing was actually shipped, or overstated shipping charges relative to the product and mode of shipping. It is not uncommon to find invoices that contain shipping charges when the order was for service, support, and software and licenses that were downloaded or e-mailed. Cross off the charge and short pay the invoice by the shipping amount. If the freight charges seem out of range relative to the product and shipping method, ask the vendor to forward the actual shipping bills from the shipper for the last 5 invoices and compare against what you were charged. Demand an immediate refund or credit for any discrepancies, and require the vendor to agree to spot audits quarterly for a comparison of charges.

Tax charge errors. Similar to shipping charge errors, tax errors typically fall into two categories; sales tax charged on a non-taxable item, or incorrect tax rates charged. Many vendors will charge tax on items that are actually non-taxable. While what is taxable differs from state to state, frequently vendors charge taxes on services, shipping, and certain types of support when those items should be tax-free. Vendors also frequently charge the incorrect tax rate on some or all line items because they fail to differentiate between products and services that may have different rates in your locality, or they apply the wrong locality's rates. Before looking for these errors, briefly familiarize yourself with the applicable tax codes and rates. If tax errors are discovered, short pay the invoice when sending payment or demand an immediate refund or credit if the bill has already been paid.

Interest charges. Many vendors will tack on a late fee or interest charge for overdue balances. Unless specifically agreed to in your terms, ensure these fees are deducted as well before paying the invoice.

Addressing the Problem

Organizations should take several proactive steps to ensure that vendor invoice errors are caught and that appropriate steps are taken to short pay and receive credits for these errors.

First and most importantly, make sure your organization avoids overpayment caused by these errors by improving your invoice approval process. Many organizations do not utilize a process that formally and systematically reconciles an invoice against the original order, instead leaving the effort undefined and manually reviewed. An optimal invoice payment process includes the use of a system that prohibits the payment of unordered line items and quantities. In fact, many organizations that implement a formal invoice reconciliation application find that their payables workload increases as they uncover and attempt to resolve the errors they now frequently discover. The lack of this effort before implementation represents clear evidence that most errors previously went undetected costing the organization in hard dollars.

Second, an audit should be performed to uncover invoice errors in the past few months. When errors are uncovered, demand an immediate refund or credit from the vendor(s). To increase the value of the audit and minimize the audit effort, employ the following techniques:

- Only go back 3-6 months; after 6 months vendors are typically unwilling to reimburse for their mistakes without additional pressures and/or threats of legal action.
- Sort your invoices by dollar size and pull the largest for review.
- Sort your invoices by number of line items and pull the largest for review.
- Sort by product/service type and pull those with substantial support and service components for review.

With this approach, an organization can maximize its return on the investment in audit resources. Organizations may also wish to leverage the services of an experienced external audit/recovery firm that will work on a contingency basis receiving only a percentage of any actual amounts recovered.

Summary

Don't let a vendor's poor invoicing processes result in higher costs for your organization. Implement an invoice reconciliation system to catch vendor invoice errors and audit the last 3-6 months of invoices for errors and possible refund opportunities.

28

Understanding and Leveraging a Vendor's Comp Plan

For some IT products or brands, understanding how a vendor compensates its sales force is perhaps second only to competition in achieving the best pricing. By understanding how your vendor representatives are compensated, and hence, motivated, organizations can potentially control their behavior and actions more than their own company can. Consider a number of factors that impact a rep's compensation plan:

Deal Timing. Best pricing is typically achieved at the end of the vendor's fiscal quarter (EOQ). But do you know whether your rep wants the deal to come in this quarter, or the next? If he has already 'hit' his number this quarter, he may actually want to 'sandbag' the deal until the next quarter, regardless of what his own company wants. Signs usually include an aggressive rep that suddenly begins to visit or call less frequently as the EOQ approaches, or is not as aggressive in pricing as they typically would be given the EOQ timing. To ensure their motivation is consistent with your timing, position larger deals as a 'use or lose' transaction; the deal gets done or the funds will no longer be available.

The Quota. Understanding what quota a vendor rep has, and where he is against it can provide significant leverage. Will this deal help him or her make their numbers for the quarter? For the year? Or is it a drop in the bucket? Knowing how important the deal is to a rep personally can help you position the purchase in a way that motivates them to find additional discounts or value for your organization.

The Overall Comp Plan. What is the rep's commission on their product or service? Is this deal material to the rep, regardless of the deal size? Big deals could be meaningless or small deals important based on a rep's overall comp plan. A larger order early in the year may be less important than a small one late in the year if the rep's plan includes 'kickers' or 'accelerators' that give him extra income for exceeding their quota.

The Product and Service Commission Mix. Vendors frequently compensate their reps differently on various products and services. Many vendors will utilize a mixed plan providing higher commissions on higher margin products and conversely, lower or no commissions on low margin products. Often a rep will also be incentivized by short-term bonuses to sell specific products that a vendor has had trouble selling or wants to seed in the market. A rep might become considerably more aggressive on pricing if it contains a non-material part or minor configuration change that results in a higher commission.

Other Duties. You may be able to help a rep out in other ways besides simply buying products. Knowing what kind of forecasting a rep is responsible for can make the information you have about your purchasing plans an asset. Additionally, some reps are responsible for other activities including customer support functions and minimizing the impact of these requirements may add additional motivation for some reps to further reduce pricing.

There are many methods of uncovering comp information. However, one of the quickest ways to capture this information is a simple lunch or dinner with your rep. Additionally, reps are often more than willing to discuss the compensation plans of companies other than their own, providing information they have gathered from their peers or through their interview processes.

Summary

Keep in mind the goal isn't to abuse or exploit a vendor's comp plan. However, by understanding how their compensation plan works, you will have a much greater understanding of how deal flow and economics can work to your organization's advantage.

29

Using Independent Service Organizations (ISOs)

When vendors sell a piece of hardware, they're often counting on more than the initial product margin, they're counting on a 3-5 year revenue stream from the annual support contracts or warranty uplifts that many companies buy along with the product. The gross margins on support can be as high as 100% for the vendor, depending on the customer's use of the support.

While vendors will often provide significant discounts on annual support after negotiation, Independent Service Organizations (ISOs) represent an alternative to vendor-direct support. An ISO is an entity that specializes in the repair and maintenance of computers or other equipment. Before leveraging an ISO as a support option however, there are several factors to consider when examining the viability and cost savings to an organization.

The first consideration impacting the ability to utilize an ISO is whether the vendor has implemented feature-sets that limit your ability to use support other than what the vendor provides. For instance, products that use a 'call-home' feature where the system will dial up or send out over the internet messages to the vendor's call center may make it unfeasible to use an ISO for that hardware.

Another consideration is support and resource quality relative to the vendor. However, in many cases an ISO is actually the delivery resource of the service an organization purchased from the vendor itself. In cases where an organization is aware the vendor is providing support by using an ISO, it's often worth examining the price differences between the ISO and vendor given that the organization will receive the same personnel and parts inventory either way.

Another potential benefit for using an ISO is that they frequently provide multi-vendor services, allowing an organization to aggregate its support dollars spent with various vendors to gain even greater volume discounts from a single ISO vendor.

A common strategy for many organizations is to use an ISO for support of non-production hardware, such as development and test systems. For production systems that require higher availability, the vendor support is instead purchased. At a minimum, the threat of moving some or all of your support business to an ISO frequently diffuses the vendor's feeling of monopoly over support on their own products giving an organization the leverage it needs to negotiate better discounts on support.

Summary

ISOs may represent an ideal vendor channel for lowering vendor support contract costs, especially if they are the primary resource the vendor itself uses to deliver support to its clients. However, consider the impact on administrative costs as well as the viability of using ISOs in conjunction with other contract support options. Even in cases where using ISOs is not an option, consider mentioning to vendors the idea of using ISOs to help secure the best possible price on vendor support.

30

Trade-Ins Equal Discounts - So Buy a Few

Although not heavily advertised, many hardware vendors occasionally offer significant additional discounts based on the trade-in of existing equipment. Designed to encourage earlier than planned technical refreshes or to replace competing products, vendors may provide set or negotiated discounts off of list prices on top of any already negotiated discount levels if an organization agrees to turn over equipment to them. In most cases, these promotional programs can be leveraged to provide discounts that far outweigh the value of selling old and idle assets or even the cost of acquiring systems from used hardware vendors specifically for trade-in. There are however several aspects of these trade-in programs that buyers should be aware of.

First, always arrive at an agreeable price prior to introducing the concept of trade-in products for additional discounts. Many vendors will simply blend the trade-in discount into the traditional discount resulting in a lower overall discount, if given the chance.

In some cases, it may be worth acquiring the equipment required for trade-in. Several channels exist that can supply qualifying systems and storage for what amounts to pocket change. From eBay to liquidators, servers originally selling for $50,000 or more are available for $400. Those same servers may provide an additional 2-10% off of the list price of a $250,000 server.

Another important concept to remember is that many of these programs have enormous flexibility in what qualifies for trade-in. Not long ago one hardware company offered a trade-in discount on a new server if two existing 'servers' were traded in. However, their definition of a server included systems then selling for less than $100. The result was a potential additional discount of more than $10,000 on a midrange system in exchange for hardware valued at less than $200. Such discounts are made possible because for the most part, unless the actual resale value of your trade-in is substantial, vendors are concentrating on overall deal economics, what they consider an acceptable gross margin, and future technology lock-in, not the value of the equipment traded in.

Vendors also frequently do a poor job of handling the logistics of trade-ins. Some never forward the required documentation and shipping instructions, others fail to follow-up to ensure equipment has been returned, and at least one has asked that the client merely throw away the equipment rather than forwarding it to them.

When contemplating a trade-in of equipment with material value, check to see what the going liquidation value is relative to the discount and act accordingly. Given that many IT assets are leased, remember to verify ownership of the equipment to avoid trading in or agreeing to trade in equipment that will need to be returned to a leasing company.

Summary

Organizations should talk to their sales representative to determine whether a trade-in program exists and understand its details. Find out how the program is structured, what kinds of equipment qualify, and calculate the value of potential discounts. Next determine what equipment is available or could be acquired to trade-in to obtain the largest additional discount.

31

Identical Products, Different Labels: OEM Alternatives

Once a specific technology product is selected, most buyers immediately think about which channel will provide the best pricing. But few think about which brand. In today's environment, smart technology buyers are doing more than choosing between technologies, they're evaluating identical products from different vendors.

From hard and tape drives to server components, many vendors use another vendor's products within their own. Known as Original Equipment Manufacturer (OEM) components, some vendors are even re-labeling entire products with their own brand in order to satisfy a niche or gap in their own product lines. Several factors can help an organization decide when to evaluate OEM versions of products they intend to purchase as well as how to use their existence to obtain better pricing.

The most important factor to consider is the potential interoperability issues related to employing a different vendors' product flavor. Differences in software and drivers may make it impossible to use the same product from a different vendor. However, in many cases the product is identical in all ways except the label on the outside of the box and operability is not an issue.

Another important factor to consider is the potential for support issues related to running different brands together. For example, a vendor may insist that they will only support a system if an installed card is purchased through them and not support a system containing the exact same card that the organization obtained from someone else.

Also, as an organization increases the number of vendors in its environment, its support headaches may increase. It may be cheaper in the long run to get everything from one vendor instead of several based on the support issues an organization may encounter as well as the increased number of vendors they will have to work with. However, many organizations might buy direct from the original manufacturer to specifically gain better support than what is being supplied by the reselling vendor.

At a minimum, it does not hurt to raise the potential with vendors. The existence of OEM partners selling the same products gives an organization another way to increase its vendor universe. For example, a vendor may have given special pricing to a single preferred reseller, limiting an organization's channel choices; but bringing up OEM can open a new set of potential competitors. Bringing the other potential products to the table means the vendor will have to re-evaluate its pricing in a truly competitive environment, not simply one of its own making.

Summary

Conduct simple research online to determine if OEM competitive alternatives exist for many of the technology components your organization invests in. Even in cases where the added technical or support complexity of using OEM products makes them unviable in your environment, consider using them for competitive leverage during vendor negotiations.

32

Buying and Managing Hardware Support

In many organizations, the purchase of hardware maintenance and support contracts is not given sufficient consideration relative to product selection and price negotiation. Moreover, support levels, terms, and renewals are often ineffectively managed. Based on client organizations observed, a typical organization with an annual IT budget of ten million dollars or more can frequently reduce support contract costs by 15-35% annually without sacrificing its service levels.

Common Weaknesses

Common practices associated with hardware support contracts include:

Paying for support on decommissioned equipment. Organizations that do not employ a formal approach to asset tracking or fully leverage their asset tracking system to manage support details increase the probability of paying for support on equipment that has been removed from service. Organizations that use spreadsheets to manage support details have found it necessary to periodically reconcile their equipment against the support list, which often leads to support cancellations.

Lack of rationalization of support levels. Organizations often utilize a single support level for a given vendor's products resulting in service levels that are incongruent with actual needs. Commonly, organizations purchase 'Platinum' or 'Premium' support for all products, including those that may be utilized in development or test environments that do not require high availability. The lack of support rationalization often stems from inattention to support details, lack of support planning, or a desire to minimize or consolidate the number of support contracts managed.

Not using spares. Many organizations do not maintain any type of sparing policy instead relying on support contracts to supply all required parts. Failing to leverage the intellectual equity of on-site staff, organizations may end up paying the equivalent of a new system every year in the form of support payments. In many cases, investing in spare parts or spare systems is significantly less expensive than the cost of vendor support contracts over the long run.

Not taking advantage of all possible support discounts. Most organizations fail to assess the terms and discount structures of vendor support. Failure to identify enterprise-wide volume to obtain better overall discounts or not taking advantage of multi- year and internal help desk discounts results in higher than necessary support contract spend.

Lack of competitive negotiation. Many organizations fail to realize that support and renewal fees can be negotiated.

A lack of meaningful competition coupled with a vendor's feeling of monopoly towards support of their own products often results in lower discounts.

Poor enforcement of Service Level Agreements (SLAs). Many support contracts include specific SLAs with penalties for not meeting agreed-upon support levels. Organizations, however, may not track agreed-upon metrics and do not enforce SLA commitments.

No rationalization of third party support costs. Many organizations use fixed-price, labor-only support agreements that only outline support responsibilities. Often when these contracts are reconciled against a meaningful metric, such as cost per hour, they are not competitive with more straightforward staff augmentation rates. These contracts may also include provisions for procurement of hardware needed for repairs that circumvent the standard procurement process, resulting in higher part costs. Additionally, these contracts often fail to utilize existing warranty or vendor support features that would reduce the overall repair cost.

Best Practices

To address these issues, a set of practices in purchasing and managing vendor support has been identified. Utilizing these practices has historically resulted in a reduction of support costs:

Utilize a support/asset management system. Support contracts possess many characteristics including contract numbers, terms, equipment covered, coverage windows, response times, parts delivery response, authorized initiators, and support contact information. Proper management of each of these variables can affect both the cost and effectiveness of service delivery. Additionally, the use of a support management system will provide enterprise-wide visibility into support spending that can be leveraged for better discounts. Support management systems should also be integrated with purchasing, accounting, and help desk systems to ensure data is current and that support details are properly tracked on new, existing, and decommissioned equipment.

Develop support tiers. The process of defining application availability and subsequent support levels on a system-by-system basis often proves to be logistically and administratively unrealistic. Defining a set of support tiers, assigning equipment to specific tiers, and the subsequent purchase of matching support ensures both manageable and cost effective support purchases. Top organizations typically identify 3-5 support levels based on their availability requirements. Examples include:

- Mission Critical: 24x7, 2 Hour Response
- Major: 12x5, 4 Hour Response
- Minor: 8x5, Next Day Response
- Everything else

Systems, along with their supporting infrastructure are assigned to specific tiers, and the minimum level of support is purchased that meets the tier's availability requirements. This method of evaluating support requirements facilitates selection and management of support levels, and maximizes the cost benefit of matching ordered support to a system's actual availability requirements.

Utilize spares where appropriate. Support needs typically exist that may make the use of on-site spares more effective than purchased support. If the ability and know-how exist within the organization's IT department to effect repairs, the cost of spare parts can be dramatically lower than the purchase of a full-service support contract. In some cases, the use of in- house skills and spare parts can also enhance overall service levels through reduced response and repair times.

Leverage all possible discounts. Review support contract terms and options and if possible, take advantage of additional discounts for terms such as multi-year commitments, or providing first-call help desk capabilities. For many vendors the termination conditions do not change, thus organizations can take advantage of these discounts without giving up flexibility. However, organizations don't typically benefit from pre-paying more than one year of support. The typical rate of return achieved from a pre-pay discount is usually less compelling than other potential investments. Additionally, vendors often complicate the process of obtaining refunds for cancelled support.

Negotiate support in a manner similar to hardware and software purchases. Introduce competition and review third party offerings from independent service providers to extinguish vendor monopolies on support of their product. Ask that support quotes be separated from hardware and software purchases and negotiate their pricing separately. In addition, ask for free support uplifts, or warranty extensions when making new product purchases.

Measure actual support and service delivery against SLAs. Develop or leverage existing metrics and reporting from help desk or related systems and obtain agreement from vendors for their use in enforcing SLAs. Measure service delivery metrics against SLAs, and request appropriate credits when service levels are not met.

Rationalize total support costs. When fixed-price service-only contracts are utilized in areas such as break/fix or desktop support, measure the cost per ticket, per seat, or per hour to determine the cost effectiveness of the contract price against alternatives, such as employing more internal resources or using Time and Materials (T&M) relationships. Additionally, ensure that products purchased to address repair needs are approved and procured in the same manner products are traditionally purchased and are not provided by the supporting organization at higher prices.

Summary

By employing better support purchasing and management practices, organizations can typically save 15-35% of their annual IT support costs. Additionally, these practices are relatively straightforward to implement and their low impact on other corporate processes coupled with minimal integration costs allow organizations to immediately reduce their support budgets.

33

User Delivery Confirmation: Make it Better and Do More

For most organizations, going through the process of confirming product or service delivery internally with users works one of two ways; poorly or not at all. Before examining the major reasons why, and how to improve the process, let's look at why delivery confirmation is beneficial. The primary benefits of a formal user delivery confirmation process include ensuring:

- The correct products and quantities are received.
- Goods are received in the expected condition.
- The services expected have been performed.

This information is critical to ensuring that invoices are paid (or short-paid) appropriately and that any issues with an order are identified and resolved as quickly as possible. However, delivery confirmation processes are often abandoned early on in their implementation or at best, utilized only when an end user decides to follow it. There are typically two major reasons that lead to this breakdown.

First, there's usually no consistent trigger for confirmation. Someone has to remember to send a request for delivery confirmation, and there is typically no consistent trigger to initiate it being sent. Many organizations use receipt of the vendor invoice as a trigger but that's typically too late for the data collected to be of any value; previously unidentified issues may be weeks old by the time an invoice is received.

Second, there is no motivation for the end-user to be bothered to complete confirmation. Once an end-user has received his products or had services performed, there is little reason for him or her to fill out a form related to the purchase. The end-user simply cannot see any direct benefit to him or her or their department.

Solving these issues is critical to effective management of the purchasing and payables cycle. And the only truly effective method is to automate the process. An organization's purchasing system must be configured or augmented to manage the workflow by sending e-mails until all parties have responded with the appropriate data. The most critical components are:

Collecting shipping information from the vendor. Most purchasing systems capture a 'need by' date. Using this field as a trigger, have the system send an e-mail to the vendor requesting shipping information. The e-mail can contain a link to a web-based form where the vendor can enter the estimated or actual ship date and tracking information. The form results can be entered directly into a database, which then triggers future e-mails to the vendor for shipping info if the date is an estimated one or to the end user if actual shipping information is provided.

Don't let end-users off the hook. Configure your system to ping end-users every few days via email until they return the required information. Be sure the mail includes a brief standardized statement explaining why the information is required (not optional) to minimize the number and frequency of complaints related to the requests. Let them know that providing the information reduces the overall spend of their group by eliminating invoicing errors.

Of course, while you're asking the end user to confirm products and quantities, you can also use the opportunity to gather other information as well, including:

Asset data. Delivery confirmation is the perfect time to collect data for your asset management process. Your confirmation form could collect serial numbers and location information at the quantity level. This data can be used to populate asset management systems or leveraged by help desk systems.

Vendor and purchasing feedback. Did the product arrive when and as expected? Was the price as expected? Include some standard questions as well as an open field for feedback. This data will not only allow you to discover unmet end-user needs but can provide you with additional ammunition when you negotiate pricing or other terms with the vendor. Make sure the relevant results also get delivered to the vendor as well.

Summary

User delivery confirmation, even if performed by user support personnel who install or configure products, is a critical component for ensuring order accuracy, catching invoice errors, and collecting additional information useful for vendor and user management.

34

Protecting Yourself with True-Up Clauses

Once the domain of software subscription contracts, more and more software vendors are including 'True-Up' clauses in their license contracts. A True-Up clause allows an organization to install software on as many systems as desired, reporting and paying for the number of licenses they have used on a periodic basis. The vendor then bills the client for the incremental license count.

The primary benefit of inserting such a clause into a software license agreement is to avoid the potential liability associated with piracy. Most organizations who are guilty of software piracy are so not because of a desire to steal, but because of poor license management. Software is often installed with the intent of purchasing a license later, but the follow-up never happens. Properly worded True-Up clauses can protect you from this activity.

Another benefit of True-Ups is a potential increase in the efficiency of your operations, enabling you to install needed software without waiting for the procurement cycle to deliver the product. This activity should be carefully monitored to ensure that proper approvals for purchase are received, and that asset management processes are followed.

Some vendors may require that the True-Up clause provide permission for the vendor to audit the license count. The primary downside of a vendor audit is that it provides the vendor with an opportunity to gain significant visibility into your IT organization. The audit can also cost you time and resources to facilitate.

While virtually all True-Up clauses do not refund you for licenses you de-activate, they can offer solid protection against piracy liability as well as enable more efficient operations if properly managed. As always, if a vendor is unwilling to accommodate your request, you can leverage their lack of flexibility in this area for a better discount or other concession.

Summary

Insist on true-up clauses, which enable an organization to periodically pay for additional licenses in use without penalty, as part of any enterprise-wide licensing agreement. The clause protects organizations from incurring additional expenses associated with non-compliance with negotiated terms as well as reduces risks associated with external or independent audits by the Software Publishers Association and others aimed at reducing software piracy.

35

Worst Practices: Use at Your Own Risk

Securing the best possible price on IT products is critical to the health of any organization that spends a material amount on information technology. But there are tactics both ethical and unethical that can damage not only your future ability to secure the best possible pricing but also you and your organization's reputation. Common practices that do more harm than good include:

Sharing one vendor's pricing with another. You may be able to get a slightly better price on a few deals, but eventually, no vendor will want to do business with you. The harm to a vendor is less that particular deal and more that you are providing their competition with a critical example of how they price and present offers. Never do it.

Putting a vendor through too many fire drills. Needing a price now and needing the product tomorrow gives a vendor a chance to shine, but eventually, the cost of accommodating your recurring emergencies shows up in your price, as the vendor attempts to adjust his profit based on their level of effort.

Making vendors pay for every mistake. Asking your vendor to eat every mistake, theirs and yours will first cost you higher prices on future orders, and eventually cost you the relationship.

Using a vendor exclusively to get a quote to help beat down the price of your first choice. Vendors are willing and more than happy to provide pre-sales support and prepare price quotes. However, if your sole goal is to simply leverage the quote in order to obtain better pricing from the vendor you've already decided to go with, think again. Vendors catch on quick and using a vendor exclusively for this purpose can result in reduced interest from other vendors reducing your long-term competitive leverage or come back as higher pricing when you really need the vendor for something.

Using RFPs and ignoring the responses, changing the requirements, or awarding business on different criteria. With vendors exhausting considerable resources responding to RFPs, it is the responsibility of the issuing organization to make sure that it legitimately plans to evaluate and award business based on them. If vendors feel typical RFP standards are not used in awarding business, vendors will likely invest fewer resources or ignore all together future RFPs. Also, when an organization has crafted an RFP so that only one vendor's product meets the requirements, essentially resulting in a 'sole-source' RFP, that organization is weakening vendor interest in future requests. If a specific product is desired, but internal processes dictate an RFP, solve the issue internally rather than burdening vendors with useless efforts.

Summary

These are just a few of the more common practices that hurt an organizations reputation and buying power. Regardless of size or industry, fair and straight-forward business practices will always be a cornerstone of an effective IT procurement function.

CLOSING THOUGHTS

Cutting IT costs is truly part art, part science. While many great ideas for reducing IT spending exist, and this book covers only a handful, ultimately the ability to increase efficiency is typically based on five fundamental factors.

Leadership's Level of Passion and Commitment to Efficiency

When leadership is apathetic about efficiency, the organization as a whole is typically apathetic as well. This lack of passion is often ok, especially in businesses with substantial product or service margins where time and energy may be best spent on areas other than squeezing every last nickel out of operations and administrative dollars.

However, few businesses have this luxury. In most organizations the leadership's commitment and passion about IT efficiency acts as a driving force that pervades over time. As an extension of the concept, 'what gets measured gets done', most organizations eventually march to the beat of its leadership drum, even in cases where initial resistance exists. As long as the emphasis is consistent, the organization's mindset, and with it each individual's efforts, will focus on efficiency as a key goal of every IT initiative or activity.

Treating IT Efficiency as a Continuous Operational Objective

Another fundamental factor, consistent with leadership's commitment, is treating IT efficiency as an enduring concept embedded in every service and activity an IT organization engages in. In many organizations, IT cost cutting is viewed as a periodic event, with sporadic spot audits or spending reviews that uncover a handful of opportunities for additional efficiencies.

These efforts are not without merit. However, they often have little sustainable impact because the mindset is not embedded in the culture, resulting at best in some level short-term savings followed by more spending on 'bad costs'. When treated as a continuous objective, efficiency becomes a permanent management concept. Like other measures of performance including benefits realization, risk management, quality, and delivery schedules, efficiency becomes a tenet of operational performance.

The Organization's Appetite for Change

As illustrated in the efficiency phase model, fundamental efficiencies in IT spending frequently require a particular operational approach. Even in federated models, the levers of efficiency that rely on centralization exist. However, organizations that operate in a mostly decentralized model, or one with few controls, may require substantial change to realize material reductions in IT spending.

But change is hard. Organizations and the people within them frequently must relinquish some level of control. People may be re-assigned or even let go to realize efficiencies. Previously accepted requirements are reconsidered. In many cases, strong leadership, good planning, and sound communication may simply be outgunned by culture, existing constituencies, and political maneuvering. In these cases, the ability to apply consistent pressure and accept incremental progress over time is often the required strategy for overcoming all too common obstacles to change.

An Idea's Fit Within a Given Environment

Every environment is unique and a one-size-fits all approach simply doesn't exist in virtually any aspect of technology implementation, operations, and management. What works in one organization may not work in another, even in what may seem a similar environment. In one organization for example, the procurement organization essentially represented the 800 pound gorilla and no business case, no matter how sound, was going to wrest any acquisition control from the group. Many otherwise obvious opportunities remained unrealized based on, from the IT organization's perspective, an intractable obstacle to progress.

An organization's culture, historic evolution, even vendor and technology footprint can increase or decrease the feasibility of short, medium, or larger-scale efficiency opportunities. The frequent key to successful cost take out or avoidance is understanding the organizational landscape and using it to inform opportunity selection.

The Business and Technical Acumen of IT Managers

Ultimately, individual project and organizational management and technical capability impact efficiency significantly. Seemingly great project managers deliver on time and budget useless technology because they lacked the technical or domain expertise to recognize during the effort that it would not deliver its intended value. IT managers fail to differentiate between projects and continuous operational activities, often investing significant amounts to support short-term efforts or failing to invest enough in supporting or automating an activity that will endure perpetually.

Individuals who lack real IT management and technical project expertise frequently represent huge buckets of organizational waste. Millions are wasted in some IT organizations annually simply because the broader leadership fails to recognize the value of strong and truly experienced IT management capability. Organizations that pay for quality, whether its internal or external, often see the additional investment returned in the form of spending avoidance that is exponentially greater than the investment.

Summary

Organizations with strong leadership commitment to efficiency, an appetite for change, and experienced management are well positioned to realize or continue realizing substantial efficiencies in IT spending. However, even those lacking these fundamental factors can enjoy, through a wide-variety of operational or tactical opportunities, reductions in or avoidance of unnecessary spending while working to incrementally change the broader IT organization towards a larger, more permanent model of efficiency.

About the Author

As an operational leader, management consultant and trusted advisor for over 18 years, Patrick Chapman has helped dozens of public and private sector organizations improve performance, identify and reduce costs, and realize true value from change. Over the course of his career, Mr. Chapman has primarily focused on helping organizations improve the efficiency and effectiveness of their information technology investments and operations. His deep technical and management expertise spans governance and steering; technical business management; systems design and development; infrastructure architecture and operations; and seat management.

In addition to his technology experience, Mr. Chapman has also helped clients improve programmatic performance as well as navigate and manage large-scale change in complex, political environments. Mr. Chapman's previous book, 'Getting Things Done in Government', focuses on helping individuals overcome highly political and decentralized operating cultures in order to drive change.

Mr. Chapman earned his bachelor's degree in Business Administration from California State University, Northridge.

Linkedin Profile:
www.linkedin.com/pub/patrick-chapman/9/a8a/b56/

Printed in Dunstable, United Kingdom

91219596R00097